Current Concepts in Uveal Melanoma

Developments in Ophthalmology

Vol. 49

Series Editor

F. Bandello Milan

Current Concepts in Uveal Melanoma

Volume Editors

Martine J. Jager Leiden
Laurence Desjardins Paris
Tero Kivelä Helsinki
Bertil E. Damato Liverpool

37 figures, 23 in color, and 12 tables, 2012

KARGER

Basel · Freiburg · Paris · London · New York · New Delhi · Bangkok · Beijing · Tokyo · Kuala Lumpur · Singapore · Sydney

Martine J. Jager
Department of Ophthalmology
Leiden University Medical Centre (LUMC)
PO Box 9600
NL–2300 RC Leiden (The Netherlands)

Laurence Desjardins
Department of Ophthalmology
Institut Curie
26, rue d'Ulm
FR–75005 Paris (France)

Tero Kivelä
Department of Ophthalmology
Helsinki University Central Hospital
Haartmaninkatu 4 C, PL 220
FI–00029 HUS, Helsinki (Finland)

Bertil E. Damato
Ocular Oncology Service
Royal Liverpool University Hospital
Prescot Street
Liverpool L7 8XP (UK)

This book was supported by Eckert & Ziegler

Library of Congress Cataloging-in-Publication Data

Current concepts in uveal melanoma / volume editors, Martine J. Jager ...
[et al.].
p. ; cm. -- (Developments in ophthalmology, ISSN 0250-3751 ; v. 49)
Includes bibliographical references and index.
ISBN 978-3-8055-9790-6 (hard cover : alk. paper) -- ISBN 978-3-8055-9791-3
(e-ISBN)
I. Jager, Martine. II. Series: Developments in ophthalmology ; v. 49.
0250-3751
[DNLM: 1. Melanoma--therapy. 2. Uveal Neoplasms--therapy. 3.
Melanoma--diagnosis. 4. Uveal Neoplasms--diagnosis. W1 DE998NG v.49 2012 /
WW 240]

616.99'477--dc23

2011033492

Bibliographic Indices. This publication is listed in bibliographic services, including Current Contents® and Index Medicus.

www.karger.com
Printed in Switzerland on acid-free paper by Reinhardt Druck, Basel
ISSN 0250–3751
ISBN 978–3–8055–9790–6
e-ISBN 978–3–8055–9791–3

Contents

List of Contributors

Bernard Asselain
Department of Biostatistics
26, rue d'Ulm
FR–75005 Paris (France)
E-Mail bernard.asselain@curie.net

Jacobus J. Bosch
Department of Internal Medicine 5,
Hematology and Medical Oncology
University Hospital Erlangen
DE–91054 Erlangen (Germany)
E-Mail jacobus.bosch@uk-erlangen.de

Nathalie Cassoux
Department of Ophthalmology
Institut Curie
26, rue d'Ulm
FR–75005 Paris (France)
E-Mail nathalie.cassoux@curie.net

Sarah E. Coupland
Department of Cellular and Molecular Pathology
University of Liverpool, 5th Floor Duncan Building
Daulby Street, Liverpool L69 3GA (UK)
E-Mail s.e.coupland@liverpool.ac.uk

Jérôme Couturier
Department of Genetics
Institut Curie – Hôpital
26, rue d'Ulm
FR–75248 Paris (France)
E-Mail jerome.couturier@curie.net

Bertil E. Damato
Ocular Oncology Service
Royal Liverpool University Hospital
Prescot Street
Liverpool L7 8XP (UK)
E-Mail bertil@damato.co.uk

Sabine Delacroix
Department of Biophysics
Institut Curie
26, rue d'Ulm
FR–75005 Paris (France)
E-Mail sabine.delacroix@curie.net

Remi Dendale
Department of Radiotherapy
Institut Curie
26, rue d'Ulm
FR–75005 Paris (France)
E-Mail remi.dendale@curie.net

Laurence Desjardins
Department of Ophthalmology
Institut Curie
26, rue d'Ulm
FR–75005 Paris (France)
E-Mail laurence.desjardins@curie.net

Mariam el Filali
Department of Ophthalmology
Leiden University Medical Centre (LUMC)
PO Box 9600
NL–2300 RC Leiden (The Netherlands)
E-Mail m.el_filali@lumc.nl

Martine J. Jager
Department of Ophthalmology
Leiden University Medical Centre (LUMC)
PO Box 9600
NL–2300 RC Leiden (The Netherlands)
E-Mail m.j.jager@lumc.nl

Tero Kivelä
Department of Ophthalmology
Helsinki University Central Hospital
Haartmaninkatu 4 C, PL 220
FI–00029 HUS, Helsinki (Finland)
E-Mail tero.kivela@helsinki.fi

Christine Levy-Gabriel
Department of Ophthalmology
Institut Curie
26, rue d'Ulm
FR–75005 Paris (France)
E-Mail christine.levy@curie.net

Livia Lumbroso-Le Rouic
Department of Ophthalmology
Institut Curie
26, rue d'Ulm
FR–75005 Paris (France)
E-Mail livia.lumbroso-lerouic@curie.net

Gregorius P.M. Luyten
Department of Ophthalmology
Leiden University Medical Centre (LUMC)
PO Box 9600
NL–2300 RC Leiden (The Netherlands)
E-Mail g.p.m.luyten@lumc.nl

Pascale Mariani
Department of Surgery
Institut Curie
26, rue d'Ulm
FR–75248 Paris (France)
E-Mail pascale.mariani@curie.net

Alexandro Mazal
Department of Biophysics
Institut Curie
26, rue d'Ulm
FR–75005 Paris (France)
E-Mail alexandro.mazal@curie.net

Edoardo Midena
Department of Ophthalmology
University of Padova
Via Giustiniani 2
IT–35128 Padova (Italy)
E-Mail edoardo.midena@unipd.it

Raffaele Parrozzani
Fondazione GB Bietti per l'Oftalmologia
IRCCS
Via Livenza 3
IT–00198 Rome (Italy)
E-Mail parrozzani@libero.it

Jacob Pe'er
Department of Ophthalmology
Hadassah-Hebrew University Medical Center
PO Box 12000
Jerusalem 91120 (Israel)
E-Mail peer@hadassah.org.il

Sophie Piperno-Neumann
Department of Medical Oncology
Institut Curie
26, rue d'Ulm
FR–75248 Paris (France)
E-Mail sophie.piperno-neumann@curie.net

Corine Plancher
Department of Biostatistics
26, rue d'Ulm
FR–75005 Paris (France)
E-Mail corine.plancher@curie.net

Simon Saule
Institut Curie
Centre Universitaire
Bâtiment 110
FR–91405 Orsay (France)
E-Mail simon.saule@curie.u-psud.fr

Xavier Sastre
Department of Pathology
Institut Curie
26, rue d'Ulm
FR–75005 Paris (France)
E-Mail xavier.sastre@curie.net

Vincent Servois
Department of Radiology
Institut Curie
26, rue d'Ulm
FR–75248 Paris (France)
E-Mail vincent.servois@curie.net

Pieter A. van der Velden
Department of Ophthalmology
Leiden University Medical Centre (LUMC)
PO Box 9600
NL–2300 RC Leiden (The Netherlands)
E-Mail velden@lumc.nl

Martin Zehetmayer
Department of Ophthalmology and Optometry
Medical University of Vienna
Waehringer Guertel 18–20
AT–1090 Vienna (Austria)
E-Mail martin.zehetmayer@meduniwien.ac.at

Preface

Every ophthalmologist will at some time encounter patients with uveal melanoma. Although it is a rare disease, the implications of having this disease are huge: not only is vision threatened, but the patient is also faced with the fact that this is a potentially fatal disease. Any resident in ophthalmology must be aware of any therapeutic opportunities, as well as possible complications.

This book provides the latest information on the diagnosis of uveal melanomas and on the wide range of treatment options. For ophthalmologists with an interest in ocular oncology, the pros and cons of each modality are described by leading experts in the field. Information is also provided on the treatment of metastases, and promising new developments that may help patients in the future. This book is intended for all ophthalmologists and other specialists who see patients with uveal melanoma, and the variety of themes will offer something new to all.

Martine J. Jager, Leiden
Laurence Desjardins, Paris
Tero Kivelä, Helsinki
Bertil E. Damato, Liverpool

Jager MJ, Desjardins L, Kivelä T, Damato BE (eds): Current Concepts in Uveal Melanoma.
Dev Ophthalmol. Basel, Karger, 2012, vol 49, pp 1–15

Diagnosis of Uveal Melanoma

Tero Kivelä

Ocular Oncology Service, Department of Ophthalmology, Helsinki University Central Hospital, Helsinki, Finland

Abstract

The diagnosis of uveal melanoma is based on clinical examination with the slit lamp and indirect ophthalmoscope together with ultrasonography of the eye. Large to medium-sized melanomas are reliably diagnosed using these methods. The challenge lies in early detection. Small melanomas are more difficult to tell from presumed naevi. A useful mnemonic 'to find small ocular melanomas' reminds the general ophthalmologist to look for tumour thickness of more than 2 mm, subretinal fluid, visual symptoms, orange pigment and location of the tumour margin at the optic disc. Optical coherence tomography and fundus autofluorescence imaging help in identifying subretinal fluid and orange pigment and in measuring the thickness of thin choroidal tumours. Each of the risk characteristics roughly doubles the likelihood of growth so that the risk for growth is about 30 times higher when all five characteristics are present as compared to their absence. In addition, a low acoustic profile, the absence of a halo around the tumour and the absence of drusen over it increase the likelihood of growth. Patients with a choroidal melanocytic tumour with at least one risk characteristic benefit from referral to an ocular oncologist. We recommend that the rest of the patients be made aware of their presumed naevus and that they should be observed periodically. The patients should also be told to return immediately if they develop new visual symptoms. Finally, the trend is toward taking a biopsy of suspicious small choroidal tumours as an alternative to documenting growth before treating them as melanomas.

The diagnosis of uveal melanoma is based on clinical examination with the slit lamp and indirect ophthalmoscope together with ultrasonography of the eye. Iris melanomas are readily observable, whereas ciliary body melanomas are difficult to detect when small because they are hidden behind the iris. Choroidal melanomas can also be missed unless the fundus is meticulously examined after full pupillary dilatation. Digital photography is most useful for documenting their size and location. When a choroidal tumour is large or peripheral, wide-angle cameras are especially helpful (fig. 1a–d).

Smaller choroidal melanomas are flat to dome in shape (fig. 1a). Exudative retinal detachment develops early (fig. 1b) and may eventually hide the tumour. With time tumours break through Bruch's membrane and acquire an essentially pathognomonic

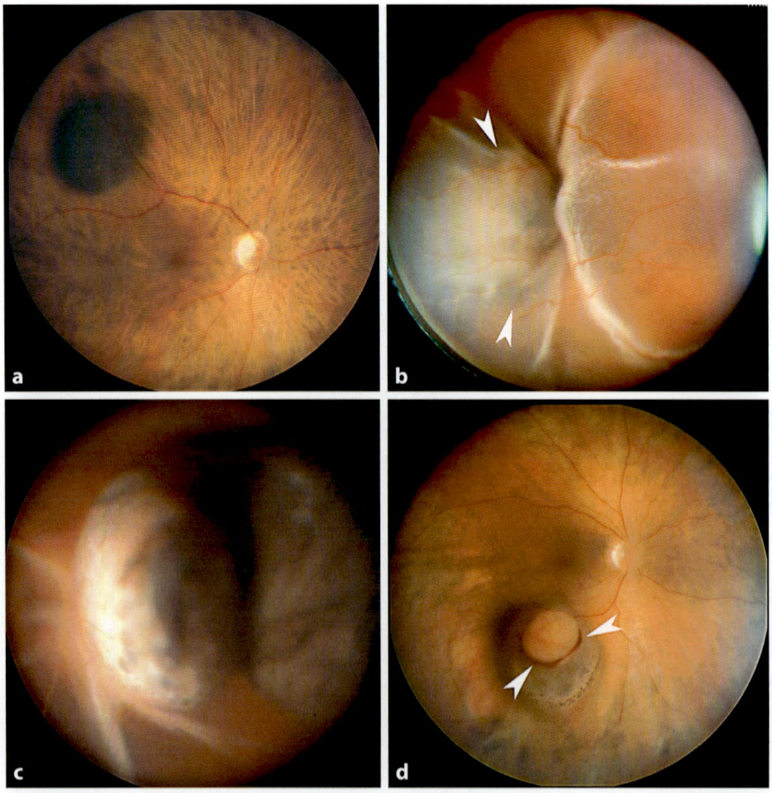

Fig. 1. Progression of uveal melanoma. **a** Small tumour that is only slightly elevated. **b** Exudative retinal detachment nearly hides a dome-shaped tumour (arrowheads). **c** Typical collar button or mushroom shape with broken Bruch's membrane. **d** Preretinal bleeding around the button after infiltrating through the retina (arrowheads).

mushroom or collar button shape (fig. 1c). If the tumour infiltrates additionally through the retina, vitreous bleeding characteristically develops (fig. 1d).

Uveal melanomas display classically low reflectivity in A- and B-scan ultrasonography. In an A-scan, the spikes are initially high but steadily diminish toward the sclera. In a B-scan, the tumour appears dark and acoustically hollow (fig. 2a). The reflectivity is less than that of the surrounding choroid, which gives the impression of a choroidal excavation at the base of the tumour. When Bruch's membrane is broken, the characteristic mushroom shape is easily observed. Blood flow within larger intratumoural blood vessels may be visible as shifting echoes. Ultrasonography is also sensitive in identifying extraocular growth, which likewise appears as a low reflective area as compared to the normal orbital tissues.

In the special case of ciliary body melanomas, high-frequency ultrasonography or ultrasound biomicroscopy is most useful, allowing visualization of tumours which otherwise would be hidden. Anterior segment optical coherence tomography is a

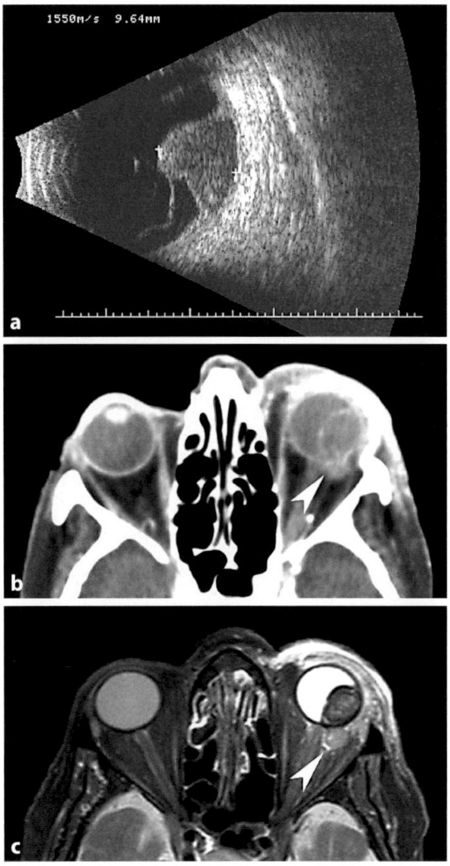

Fig. 2. Imaging of uveal melanoma. **a** The acoustic reflectivity of uveal melanoma is low resulting in a dark or 'hollow' appearance on B-scan. Collar button shape is obvious. **b** Signal intensity of viable melanoma, here extraocular (arrowhead), is higher than that of the vitreous in T_1-weighted magnetic resonance images; the intraocular part is isointense because of necrosis. **c** In T_2-weighted images, signal intensity of viable melanoma is lower than that of the vitreous.

newer technique for visualizing iris and ciliary body tumours, but so far its penetration is inferior to what is achieved with ultrasonography [1]. In the absence of such equipment, transpupillary transillumination, gonioscopy or oblique biomicroscopy when the patient is looking to the direction of the tumour may help to reveal a ciliary body melanoma.

Uveal melanomas have an intrinsic circulation, additional to the normal choroidal one. Demonstration of such double circulation or leakage of tumour blood vessels is today rarely needed to make the diagnosis of larger melanomas. Fluorescein angiography is mainly used when uveal melanoma is a viable differential diagnostic option but another lesion, such as haemorrhagic macular degeneration or a choroidal haemangioma, is a likely alternative. Detection of looping vessels by indocyanine green angiography may in some cases help to predict growth of small pigmented melanocytic tumours [2, 3].

Computed tomography and magnetic resonance imaging are helpful in eyes which have secondary vitreous haemorrhage, extensive retinal detachment or cataract which make visualization of the fundus difficult or impossible. They are also useful for charting more advanced extrascleral growth (fig. 2b, c). The tumour typically has high signal intensity in T_1-weighted images and low signal intensity in T_2-weighted ones, but either haemorrhage or necrosis may alter this pattern. These imaging methods are generally not necessary to establish the diagnosis of uveal melanoma, but they can be required for planning of proton beam or stereotactic radiotherapy.

The role of fundus autofluorescence imaging and optical coherence tomography in diagnosing especially small and posterior uveal melanomas is described later.

Diagnosing Large to Medium-Sized Uveal Melanomas

In the Collaborative Ocular Melanoma Study, which was launched in 1986 and closed for accrual in 1998, medium-sized choroidal melanomas were randomized to iodine plaque brachytherapy or enucleation, and large ones to enucleation with or without pre-enucleation external beam radiotherapy [4]. Altogether, 660 medium-sized and 1,003 large presumed choroidal melanomas were randomized to enucleation. The participating clinicians had correctly diagnosed 99.7% of the tumours enucleated [5].

The Collaborative Ocular Melanoma Study trials tell us that an *experienced retinal specialist or ocular oncologist* will have little difficulty in diagnosing a *large to medium-sized* uveal melanoma, even when using the imaging and ultrasonographic equipment of the 1980s and 1990s, which was somewhat crude by modern standards.

In these size categories, the problem is not in diagnosing a tumour as a uveal melanoma but in finding it and in referring the patient. Studies which looked for diagnostic delays suggest that 28–37% of uveal melanomas were not found by the ophthalmologist or optometrist on the first visit although many of the tumours were symptomatic at that time [6–8]. To avoid such a delay, it is imperative to dilate the pupil and to examine the fundus with binocular indirect ophthalmoscopy of all patients who have reduced visual acuity or symptoms suggestive of a posterior segment lesion such as photopsia, floaters, metamorphopsia or a shadow in the visual field [6].

Diagnosing Small Uveal Melanomas

The case is very different with small uveal melanomas, which are often difficult to tell from benign naevi at first sight. Formerly such lesions were generally observed for growth [9–11]. However, models based on tumour doubling times now suggest that micrometastasis initiates one to several years before the diagnosis of a primary choroidal melanoma is generally made [12]. The predicted size of the primary tumour when metastasis may already commence is but 3 mm in diameter and 1.5 mm in thickness [13].

Small melanomas of this size can already harbour chromosomal and gene expression abnormalities associated with systemic metastasis [14], lending support to the theoretical model. As a result, the attitude toward potential small melanomas is changing [15] and how quickly and efficiently they will be differentiated from naevi may well turn out to be of considerable importance to some patients.

The general ophthalmologist is frequently confronted with the task of differentiating a naevus from a small melanoma. The estimated prevalence of choroidal naevi ranges widely from 3 to 20%, depending on the population and study design. The most recent estimate suggests the prevalence in Caucasians to be between 5 and 10% [16]. To make an efficient diagnosis, it is useful to be familiar both with the epidemiology of naevi and with signs that suggest growth and, consequently, malignancy.

How Often Do Choroidal Melanomas Develop from Naevi?

Of uveal melanomas diagnosed, 6–8% develop from previously identified presumed naevi [7]. The true percentage probably is double as high, because 40–60% of patients with a newly diagnosed uveal melanoma have not visited an ophthalmologist during the previous decade [9, 11, 17]. Thus, only one half of them are probably aware of the existence of a prior naevus. Moreover, naevi in the peripheral fundus may escape attention even when the eye has been examined in the past. These findings taken together, it is likely that about 20% of uveal melanomas develop from a previous naevus.

How Often Do Choroidal Naevi Progress to Melanoma?

Uveal naevi, on the other hand, seem to have a fairly low risk of progressing to melanoma during lifetime. Recently, the annual rate of malignant change of a choroidal naevus in Caucasians was estimated to be 1 in 8,845 on average, ranging from 1 in 269,565 in the youngest to 1 in 3,664 in the oldest age group [16].

These calculations can be carried a step further, bringing the focus to the patient level. Starting from a 15-year-old individual and cumulating the risk over lifetime, the risk of a choroidal naevus undergoing malignant change can be estimated to be 0.04% by the age of 40 years, 0.28% by the age of 60 and 0.78% by the age of 80 years [18]. The presumption is then that all melanomas would develop from naevi, which is unlikely. As argued, about 20% of melanomas may develop from naevi, in which case the adjusted lifetime risk would be 0.06% by the age of 60 years and would eventually approach 0.2% or 1 in 500 after the age of 80 years.

In practice, most uveal naevi will have a smaller than average risk of growth. For efficient diagnosis, the clinician has to identify those with a higher than average risk of growth. Many features predictive of growth have been identified over the course of years and are now well established [17, 19–23]. Several tumours with one or more of these features will actually already be small melanomas, some most likely arising de *de novo* and some by transformation from a prior naevus.

How to Find Small Ocular Melanomas (Using Helpful Hints Daily)?

The risk characteristics which independently predict growth of small choroidal melanocytic tumours were popularized by Shields et al. [20, 22] who coined the mnemonic TFSOM, i.e. 'to find small ocular melanomas'. The mnemonic helps the ophthalmologist to remember to look for tumour *thickness,* subretinal *fluid, symptoms, orange* pigment and tumour *margin* when evaluating small choroidal melanocytic tumours.

The same authors later added three features extending the mnemonic with 'using helpful hints daily': a low acoustic profile or *ultrasound hollowness,* absence of a *halo* around the tumour and absence of *drusen* over the tumour [24].

T – for Thickness

The thickness of a choroidal tumour can be grossly estimated by biomicroscopy or indirect ophthalmoscopy by imagining that the optic disc would be tilted upward. Tumours that appear higher than the disc, which is 1.5 mm in diameter, approach the 2-mm limit. Dioptric difference in direct ophthalmoscopy can also be used to estimate thickness. However, usually thickness is measured with A- or B-scan ultrasonography.

The B-scan becomes inaccurate when the thickness of the tumour is less than 1 mm. Such tumours located posterior to the equator can be imaged with optical coherence tomography, using the enhanced depth imaging technique in which the instrument is pushed closer to the eye than normally (fig. 3). This causes mirroring of the image but at the same time shifts the best resolution from the level of the retina to the choroid [25].

The 2-mm cut-off point is obviously an arbitrary one. Comparison of frequency distributions of tumour thicknesses indicates, however, that there will be approximately 125 naevi for every melanoma in the thickness range from 1.5 to 2 mm, but only approximately 25 naevi for every melanoma in the range of 2–2.5 mm and no more than 5 naevi for every melanoma in the range of 2.5–3 mm [26].

Also presumed naevi grow somewhat in thickness, especially during adolescence. In one large study, the average naevus was 0.3 mm thicker in the age group 21–50 years as compared to patients aged 20 years or less [27]. In the oldest age group of over 50 years, the average thickness was 0.1 mm larger than in those 21–50 years of age.

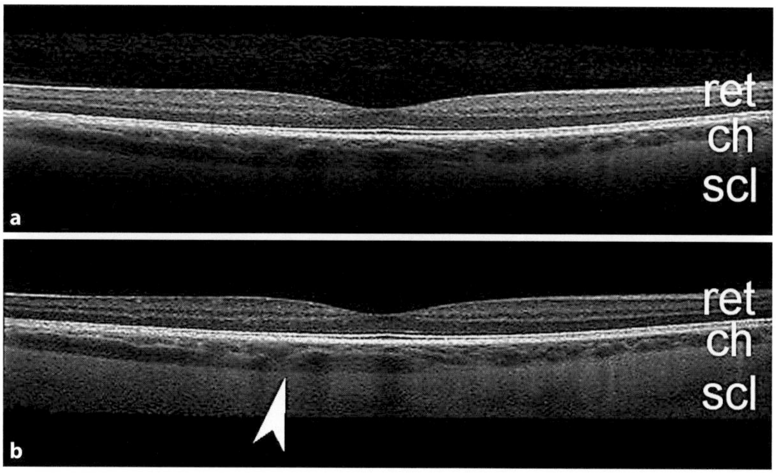

Fig. 3. a Normal optical coherence tomography images emphasize retinal (ret) detail. **b** If the instrument is moved closer to the eye, the image becomes inverted but the best resolution shifts to the choroid (ch) highlighting its margin toward the sclera (scl) (arrowhead) and enabling choroidal thickness measurements; the image shown is inverted back.

F – for Subretinal Fluid

Exudative retinal detachment is a hallmark of frank uveal melanomas [28]. Subretinal fluid at the top of or adjacent to a small pigmented tumour is felt by many to be a strong, if not the strongest, hint of a potential malignant nature of an otherwise indeterminate pigmented choroidal tumour. Small uveal melanomas may, however, lack clinically detectable subretinal fluid.

Moreover, small amounts of subretinal fluid leak from 5 to 15% of presumed naevi [27, 29], probably because of a breakdown of the retinal pigment epithelial barrier. Optical coherence tomography shows that pockets of subretinal fluid may be found over presumed choroidal naevi [30], often between larger drusen. It is not known yet whether subclinical amounts of fluid detectable over and around small pigmented tumours carry exactly the same implication as does subretinal fluid detectable clinically because studies which incriminated subretinal fluid antedate the invention of this imaging method, but preliminary evidence suggests that subclinical fluid may be relevant in this respect [31].

S – for Symptoms

Symptoms from a small choroidal melanocytic tumour are visual and are usually due to extension of the tumour or its associated subretinal fluid toward the fovea,

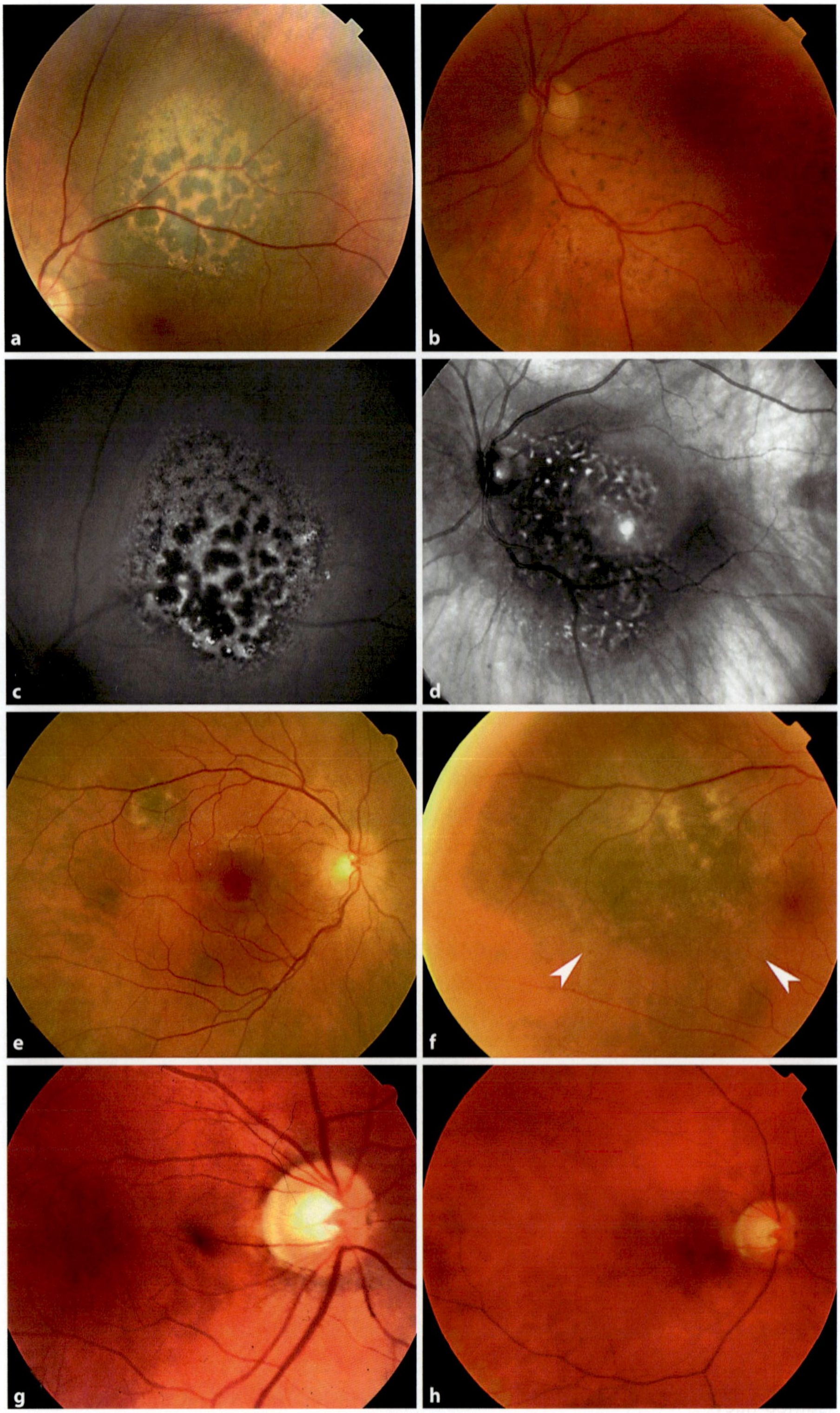
a
b
c
d
e
f
g
h

typically causing blurred vision, altered colour perception, metamorphopsia or a central shadow in the visual field [7, 20, 27].

The presence of symptoms at diagnosis often means that they have appeared recently, suggesting change in the tumour. Appearance of new symptoms during follow-up of a naevus is alarming and frequently leads to diagnosis of uveal melanoma [7].

O – for Orange Pigment

Orange flecks and subretinal fluid over a small pigmented tumour are often felt to be the strongest single indicators of potential malignancy. In fact, these two features usually occur together, because the flecks are typically located on the outer surface of the detached neuroretina. The pigment is a special type of lipofuscin, which forms especially over choroidal melanomas located in the posterior pole. It is only exceptionally detectable over more peripheral melanomas. The flecks typically are of varying size and shape and can even form a meshwork over the tumour (fig. 4a).

If the tumour is amelanotic, the pigment will appear to be black rather than orange in colour (fig. 4b). Whichever colour, it is generally brightly autofluorescent [32, 33], which has made fundus autofluorescence imaging very useful in the differential diagnosis of small uveal pigmented tumours (fig. 4c, d).

Orange pigment is also found on the surface of 6–10% of presumed benign naevi [27, 29]. Consequently, its presence alone, if somewhat alarming, does not automatically indicate a uveal melanoma. The long-term significance of orange pigment over presumed naevi has not been studied conclusively.

M – for Margin Relative to the Optic Disc

The mnemonic originally referred to tumour margin touching the optic disc as a high-risk characteristic [20, 22, 34]. In a more recent publication from the same group, this feature appears as tumour margin within 3 mm from the optic disc [24].

Fig. 4. **a** Lipofuscin over a pigmented melanoma appears orange. **b** Lipofuscin over an amelanotic melanoma appears black. **c, d** It is brightly autofluorescent in both instances. **e** Halo naevus without any risk characteristics in the eye of a 19-year-old man. **f** Choroidal melanoma with subretinal fluid (arrowheads) was diagnosed 24 years later; the halo is still seen over the tumour. **g** Small peripapillary presumed naevus without high-risk characteristics in an eye of a 54-year-old woman. **h** Five years later, the tumour is 5 times larger and has small orange pigment flecks suggesting melanoma.

According to the respective multivariate analyses, both features are significantly associated with the risk of growth, but the former one is associated with a higher relative risk.

The association between the risk for growth and the margin of the tumour touching the optic disc is likely to be a statistical rather than a biological one. In fact, it might signify that the tumour *already* has grown to reach the disc margin.

UH – for Ultrasound Hollowness

The acoustic profile of a uveal melanoma is low because these tumours are homogenous and contain few stromal elements which scatter ultrasound [35], whereas the typical choroidal naevus is equal in reflectivity to the normal choroid. Consequently, a low acoustic profile in a small melanocytic tumour suggests malignancy [24]. However, not all acoustically hollow naevi seem to grow on follow-up.

H – for Absence of a Halo around the Tumour

Some choroidal naevi are surrounded by a yellowish ring of lighter pigmentation [36]. Statistically, the presence of a halo around a small choroidal melanocytic tumour makes growth less likely [24]. About 4% of halo naevi will nevertheless eventually be diagnosed as melanomas [36] and the absence of a halo should consequently be viewed as a high-risk characteristic rather than its presence as a protective factor (fig. 4e, f).

D – for Absence of Drusen

Choroidal naevi like their cutaneous counterparts generally are present from childhood and the retinal pigment epithelium overlying them becomes compromised over time, leading to the emergence of drusen and, sometimes, atrophy and proliferation of the pigment epithelium [27]. The presence of any of these features suggests that the tumour is long standing and, by inference, likely to be benign.

However, not all naevi carry these features. Drusen are found in about 10, 40 and 60% of presumed naevi in patients aged 20 years or less, 21–50 years and over 50 years, respectively [27]. Atrophy and hyperplasia of the pigment epithelium is exceptional in patients younger than 20 years but they occur equally in up to 10% of the two older age groups [27]. The clinical impression is that the thicker the naevus is, the more likely it carries retinal pigment epithelial changes.

It is important to check whether the pigment epithelial changes are evenly distributed over the tumour. Although the presence of drusen speaks against a *de novo* melanoma, drusen do not exclude malignant change in a pre-existing naevus, especially when absent from a part of the tumour. Finally, a streak or 'tail' of retinal pigment epithelial atrophy and proliferation suggests intermittent leakage of subretinal fluid rather than just a long-standing lesion.

Additional Risk Characteristics

Regarding tumour diameter, comparison of the frequency distributions indicate that there are approximately 70 naevi for every choroidal melanoma in the diameter range of 5–6 mm as compared to approximately 10 naevi for every melanoma in the range of 6–7 mm and not more than about 3 naevi for every melanoma in the range of 7–8 mm [26].

As mentioned, looping vessels identified by indocyanine green angiography may help to predict growth of small choroidal melanocytic tumours [2, 3]. When loops were detected, about half of the tumours grew during an average follow-up of 39 months, but when loops were absent no growth was detected [3].

Observed Growth

Observed growth over a short period of, say, 1 year or less is generally taken as essentially definitive evidence of a tumour being a small melanoma. Generally, an increase of 0.3 mm in thickness and 0.5 mm in diameter as measured by ultrasonography is taken as unequivocal evidence of growth. Sequential fundus photographs may reliably identify even a smaller change in tumour diameter.

Benign naevi can also grow slowly, particularly in adolescence. In one larger study, 31% of presumed naevi grew 0.2–3.0 mm in diameter over a mean follow-up of 15 years [29]. The annual growth varied from 0.01 to 0.36 mm with a median of 0.06 mm, which translates to a 0.5-mm increase in 8 years. The percentage of growing was 54% in patients aged less than 40 years and 19% in patients aged more than 60 years. In fact, younger age was the only factor which predicted growth by multivariate analysis.

Asymmetric growth suggests malignant change of a naevus. Because a tumour will remain less than 1 mm in size for the first 20 cell divisions or so [13], it is possible for an incipient melanoma to grow hidden within a naevus for several years before reaching its margin and thus becoming identifiable.

Tumours which grow in all directions are either *de novo* melanomas, transformed naevi in which the malignant change took place centrally, or growing naevi, depending on how rapidly they enlarge. The presence or absence of high-risk characteristics will help to make the diagnosis. A *de novo* melanoma is likely to show orange pigment and subretinal fluid in the absence of drusen and retinal pigment epithelial changes, a transformed naevus may combine both sets of features and a growing naevus typically displays only degenerative features on its surface.

When and How Should One Review Choroidal Naevi to Diagnose Melanoma Early?

The presence of any one of the five original TFSOM characteristics significantly raises the likelihood of growth, which often heralds or already is a sign of malignancy, the more so when occurring together. The relative risk for a small melanocytic tumour to grow (compared with the absence of any characteristic) is 1.9 for the presence of 1 factor, 3.8 for 2 factors, 7.4 for 3 factors, 14.1 for 4 factors, and 27.1 for all 5 factors [22] Thus, the relative risk roughly doubles by every addition of one feature.

It is of note that the majority of presumed naevi that develop into a melanoma are associated with at least one of the known high-risk characteristics for growth from the beginning [7, 22, 24]. Roughly 90% of them initially had at least 1 high-risk characteristic for growth and 36% showed 3–4 such characteristics [7]. As mentioned, later onset of visual symptoms often heralds eventual diagnosis and treatment [7].

Because naevi are common the opinion on how one should proceed when one is found continues to be divided. We suggest that all patients be told about the naevus. If any of the high-risk characteristics is present, perhaps with the exception of peripapillary location of a flat naevus, it is wise to seek a second opinion from an ocular oncologist.

In the absence of high-risk characteristics, it is reasonable to propose intermittent review, the intervals of which depend both on the age of the patient and the size of the presumed naevus. If the patient is young, the naevus is small or both, the interval can be 2 years. Otherwise, we recommend initially a 1-year interval. If the patient has previously been examined without a naevus being mentioned, we start with intervals of 3–6 months. If no change is observed, the intervals can gradually be extended to 2 years. All patients should be told to return immediately if any visual symptoms develop.

We strongly suggest that the review always be based on fundus photographs taken at baseline, ideally with intermittent optical coherence tomographic or ultrasonographic measurement of thickness if the naevus is elevated. Such photographs can be extremely helpful in diagnosing melanomas when they are small (fig. 4g, h). As mentioned, optical coherence tomography is also of use in finding small amounts of subretinal fluid. Even though there is no conclusive evidence that subclinical subretinal fluid is an identical risk factor as clinically evident fluid, it can be viewed as a precursor of the latter.

In a retrospective review, the median time to diagnosis of uveal melanoma was 3 years [7]. When melanoma was diagnosed, the median thickness of the tumour had increased by 3.5 times to 3.5 mm, the diameter by 1.7 times to 10 mm and the volume by 7.5 times to 142 mm^3. The review had typically been based only on ophthalmoscopy. Had photographs been available, the diagnosis would have been made earlier in most patients.

When to Take a Biopsy to Diagnose a Melanoma?

A needle biopsy is of definite value in pursuing the diagnosis of a suspected small melanoma and in assessing its grade [37–39] although diagnostic errors remain a possibility because of heterogeneous cell lines in different parts of some tumours [40, 41], which are most likely if the tumour represents a transformed naevus. We have observed a number of naevi in which, after a stationary period of many years or decades, malignant change apparently has occurred only in a cell line on one border and it may then be tricky to get a representative biopsy sample.

Intraocular biopsy also involves the risk for vitreous haemorrhage, retinal detachment and other complications that may affect vision. Nevertheless the current trend is toward more frequent and earlier biopsy with the aim of earlier treatment of small suspicious choroidal melanocytic tumours.

Before treating a centrally located tumour, most ocular oncologists expect to see more than one high-risk sign, documented growth or a positive biopsy. Melanomas that are centrally located tend to distort vision in their early stage, which motivates the patient to opt for treatment even when it is likely to reduce vision.

If the tumour is peripheral, a suspicious lesion can be treated without definite signs of growth, for example when only one high-risk characteristic is observed, such as orange pigment or subretinal fluid. Some ocular oncologist would, however, consider taking a biopsy in order to assess prognosis.

References

1 Bianciotto C, Shields CL, Guzman JM, Romanelli-Gobbi M, Mazzuca D Jr, Green WR, Shields JA: Assessment of anterior segment tumors with ultrasound biomicroscopy versus anterior segment optical coherence tomography in 200 cases. Ophthalmology 2011;118:1297–1302.

2 Mueller AJ, Freeman WR, Schaller UC, Kampik A, Folberg R: Complex microcirculation patterns detected by confocal indocyanine green angiography predict time to growth of small choroidal melanocytic tumors: MuSIC Report II. Ophthalmology 2002;109:2207–2214.

3 Singh AD, Bena JF, Mokashi AA, Jacques R, Rundle PA, Rennie IG: Growth of small tumors. Ophthalmology 2006;113:1061–1064.

4 Singh AD, Kivelä T: The collaborative ocular melanoma study. Ophthalmol Clin North Am 2005;18: 129–142.

5 Collaborative Ocular Melanoma Study Group. Histopathologic characteristics of uveal melanomas in eyes enucleated from the Collaborative Ocular Melanoma Study. COMS report No 6. Am J Ophthalmol 1998;125:745–766.

6 Ah-Fat FG, Damato BE: Delays in the diagnosis of uveal melanoma and effect on treatment. Eye 1998; 12:781–782.

7 Eskelin S, Kivelä T: Mode of presentation and time to treatment of uveal melanoma in Finland. Br J Ophthalmol 2002;86:333–338.

8 Bove R, Char DH: Nondiagnosed uveal melanomas. Ophthalmology 2004;111:554–557.

9 Char DH: The management of small choroidal melanomas. Surv Ophthalmol 1978;22:377–386.
10 Gass JD: Observation of suspected choroidal and ciliary body melanomas for evidence of growth prior to enucleation. Ophthalmology 1980;87:523–528.
11 Augsburger JJ, Vrabec TR: Impact of delayed treatment in growing posterior uveal melanomas. Arch Ophthalmol 1993;111:1382–1386.
12 Eskelin S, Pyrhönen S, Summanen P, Hahka-Kemppinen M, Kivelä T: Tumor doubling times in metastatic malignant melanoma of the uvea: tumor progression before and after treatment. Ophthalmology 2000;107:1443–1449.
13 Eskelin S, Kivelä T: Uveal melanoma: implications of tumor doubling time (author's reply). Ophthalmology 2001;108:830–831.
14 Ghassemi F, Shields CL, Materin MA, Shields JA: Small choroidal melanoma with monosomy 3. Middle East Afr J Ophthalmol 2010;17:268–269.
15 Shields JA: Treating some small melanocytic choroidal lesions without waiting for growth. Arch Ophthalmol 2006;124:1344–1346.
16 Singh AD, Kalyani P, Topham A: Estimating the risk of malignant transformation of a choroidal nevus. Ophthalmology 2005;112:1784–1789.
17 Gass JD: Observation of suspected choroidal and ciliary body melanomas for evidence of growth prior to enucleation. Ophthalmology 1980;87:523–528.
18 Kivelä T, Eskelin S: Transformation of nevus to melanoma. Ophthalmology 2006;113:887–888.
19 Augsburger JJ, Schroeder RP, Territo C, Gamel JW, Shields JA: Clinical parameters predictive of enlargement of melanocytic choroidal lesions. Br J Ophthalmol 1989;73:911–917.
20 Shields CL, Shields JA, Kiratli H, De Potter P, Cater JR: Risk factors for growth and metastasis of small choroidal melanocytic lesions. Ophthalmology 1995;102:1351–1361.
21 The Collaborative Ocular Melanoma Study Group. Factors predictive of growth and treatment of small choroidal melanoma: COMS Report No 5. Arch Ophthalmol 1997;115:1537–1544.
22 Shields CL, Cater J, Shields JA, Singh AD, Santos MC, Carvalho C: Combination of clinical factors predictive of growth of small choroidal melanocytic tumors. Arch Ophthalmol 2000;118:360–364.
23 Singh AD, Mokashi AA, Bena JF, Jacques R, Rundle PA, Rennie IG: Small choroidal melanocytic lesions: features predictive of growth. Ophthalmology 2006; 113:1032–1039.
24 Shields CL, Furuta M, Berman EL, Zahler JD, Hoberman DM, Dinh DH, Mashayekhi A, Shields JA: Choroidal nevus transformation into melanoma: analysis of 2,514 consecutive cases. Arch Ophthalmol 2009;127:981–987.
25 Torres VL, Brugnoni N, Kaiser PK, Singh AD: Optical coherence tomography enhanced depth imaging of choroidal tumors. Am J Ophthalmol 2011;151:586–593.
26 Augsburger JJ, Correa ZM, Trichopoulos N, Shaikh A: Size overlap between benign melanocytic choroidal nevi and choroidal malignant melanomas. Invest Ophthalmol Vis Sci 2008;49:2823–2828.
27 Shields CL, Furuta M, Mashayekhi A, Berman EL, Zahler JD, Hoberman DM, Dinh DH, Shields JA: Clinical spectrum of choroidal nevi based on age at presentation in 3,422 consecutive eyes. Ophthalmology 2008;115:546–552.
28 Kivelä T, Eskelin S, Mäkitie T, Summanen P: Exudative retinal detachment from malignant uveal melanoma: predictors and prognostic significance. Invest Ophthalmol Vis Sci 2001;42:2085–2093.
29 Mashayekhi A, Siu S, Shields CL, Shields JA: Slow enlargement of choroidal nevi: a long-term follow-up study. Ophthalmology 2011;118:382–388.
30 Shields CL, Mashayekhi A, Materin MA, Luo CK, Marr BP, Demirci H, Shields JA: Optical coherence tomography of choroidal nevus in 120 patients. Retina 2005;25:243–252.
31 Espinoza G, Rosenblatt B, Harbour JW: Optical coherence tomography in the evaluation of retinal changes associated with suspicious choroidal melanocytic tumors. Am J Ophthalmol 2004;137:90–95.
32 Gunduz K, Pulido JS, Bakri SJ, Petit-Fond E: Fundus autofluorescence in choroidal melanocytic lesions. Retina 2007;27:681–687.
33 Materin MA, Raducu R, Bianciotto C, Shields CL: Fundus autofluorescence and optical coherence tomography findings in choroidal melanocytic lesions. Middle East Afr J Ophthalmol 2010;17:201–206.
34 Shields CL, Shields JA: Clinical features of small choroidal melanoma. Curr Opin Ophthalmol 2002; 13:135–141.
35 Lin DT, Munk PL, Maberley AL, Cooperberg PL, Rootman J: Ultrasonography of pathologically proved choroidal melanoma with a high-resolution small-parts scanner. Can J Ophthalmol 1987;22:161–164.
36 Shields CL, Maktabi AM, Jahnle E, Mashayekhi A, Lally SE, Shields JA: Halo nevus of the choroid in 150 patients: the 2010 Henry van Dyke Lecture. Arch Ophthalmol 2010;128:859–864.

37 Augsburger JJ, Correa ZM, Schneider S, Yassin RS, Robinson-Smith T, Ehya H, Trichopoulos N: Diagnostic transvitreal fine-needle aspiration biopsy of small melanocytic choroidal tumors in nevus versus melanoma category. Trans Am Ophthalmol Soc 2002;100:225–232.
38 Bechrakis NE, Foerster MH, Bornfeld N: Biopsy in indeterminate intraocular tumors. Ophthalmology 2002;109:235–242.
39 Shields CL, Ganguly A, Bianciotto CG, Turaka K, Tavallali A, Shields JA: Prognosis of uveal melanoma in 500 cases using genetic testing of fine-needle aspiration biopsy specimens. Ophthalmology 2011;118:396–401.
40 Mensink HW, Vaarwater J, Kilic E, Naus NC, Mooy N, Luyten G, Bruggenwirth HT, Paridaens D, de Klein A: Chromosome 3 intratumor heterogeneity in uveal melanoma. Invest Ophthalmol Vis Sci 2009;50:500–504.
41 Schoenfield L, Pettay J, Tubbs RR, Singh AD: Variation of monosomy 3 status within uveal melanoma. Arch Pathol Lab Med 2009;133:1219–1222.

Tero Kivelä, MD, FEBO
Department of Ophthalmology, Helsinki University Central Hospital
Haartmaninkatu 4 C, PL 220
FI–00029 HUS, Helsinki (Finland)
Tel. +358 9 4717 3110, E-Mail tero.kivela@helsinki.fi

Jager MJ, Desjardins L, Kivelä T, Damato BE (eds): Current Concepts in Uveal Melanoma.
Dev Ophthalmol. Basel, Karger, 2012, vol 49, pp 16–26

Treatment Selection for Uveal Melanoma

Bertil E. Damato

Ocular Oncology Service, Royal Liverpool University Hospital, Liverpool, UK

Abstract

Uveal melanomas can be treated by various forms and combinations of radiotherapy, phototherapy and local resection, which need to be tailored to the tumour size, location and extent as well as the patient's needs, fears and wishes. Many factors complicate the process of treatment selection and these include limited resources, incomplete knowledge of the patient's condition (e.g. genetic tumour type) and inadequate scientific evidence on which to base decision making. In this chapter, I describe how I select treatment for small, large, juxtapapillary, ciliary and iridal melanomas, respectively, and for melanocytic tumours of indeterminate malignancy. I also describe how I interact with patients so as to select the most suitable treatment in a personalised manner. Finally, I discuss challenges and shortcomings in this aspect of patient care, also providing suggestions for further research in this field.

The aims of ocular treatment of uveal melanoma are to prevent metastatic spread and to conserve the eye with useful vision, if possible. There are several therapeutic modalities, which can be administered individually or in various combinations to enhance the likelihood of local tumour control while minimising the risk of collateral damage to uninvolved tissues. Given any choice, treatment is selected according to the size, location and extent of the tumour as well as the patient's needs, wishes and fears. Treatment selection is complicated by constraints imposed by limited health care resources, such as unavailability of facilities for proton beam radiotherapy. There may also be restrictions arising from the patient's own circumstances, for example, travel restrictions because of the need to care for an infirm and housebound relative. Another challenge is uncertainty about the likely outcomes of rival treatments, because the required evidence is not available in the published literature or because important factors, such as histological grade and genetic type of the patient's tumour, are not known.

In this chapter, I describe a number of clinical scenarios and explain how I would select what I consider to be the most appropriate form of management. There is often a lack of consensus amongst ocular oncologists about how a particular patient or tumour should be treated. This article therefore provides my personal approach to treatment selection, with references to other points of view. I would not presume to suggest that my therapeutic choices are any better than those made elsewhere. Nevertheless, I hope that the general principles I describe will be applicable in other clinical environments.

With regard to my 'credentials' for writing this chapter, I am fortunate in having a relatively wide range of therapeutic options at my disposal, without excessive financial constraints. I am also privileged to treat patients from different parts of the world so that my experience is not limited to counselling English patients.

Clinical Scenarios

Melanocytic Choroidal Tumour of Indeterminate Malignancy

I find the management of patients with a pigmented choroidal tumour to be particularly challenging and this is because the impact of ocular treatment on these small tumours is not known [1]. It is conventional practice for these patients to be monitored until there is photographic evidence of tumour growth; however, such delayed treatment risks missing any opportunity for preventing the onset of metastatic spread in some patients. As many as 30–40% of choroidal melanomas have metastatic capability, as suggested by chromosome 3 loss and chromosome 8q gain, if treated when they are small (i.e., having a basal diameter not exceeding 10 mm) [2]. It is not known how many of these small, lethal tumours have not yet started metastasising by the time they are treated. It is also not known how many of the small, disomy 3 melanomas will transform to the high-grade, metastasising variety if left untreated for months or years until growth is seen. A recent patient of mine had a choroidal melanoma that suddenly grew rapidly after many years of apparent inactivity [3]. Examination of the enucleated eye showed the tumour to consist of spindle, disomy 3 cells basally and of epithelioid, monosomy 3 cells at its apex. This case suggests that the tumour underwent malignant transformation while the patient was being monitored. If the patient ever develops metastatic disease, there will inevitably be concerns that such a catastrophe might have been prevented by prompt treatment. Conversely, treatment of all suspicious tumours would result in unnecessary ocular morbidity and visual loss in most patients, especially as 30–40% of small choroidal melanomas extend close to the optic disc or fovea.

I approach the counselling of patients with an indeterminate melanocytic choroidal tumour in a stepwise manner. First, I explain what the ocular treatment would entail, providing an estimate of the chance of visual loss. Next, I suggest that delaying treatment until the basal tumour diameter has increased by 1 mm may increase the 10-year risk of metastatic death by around 1–2% (i.e., 1% if I consider the chances of malignancy to be fifty-fifty and 2% if I am confident that the tumour is indeed a melanoma). This is based on my observed increase in the 10-year mortality of approximately 4% and mentally adjusting for lead time bias [4]. The patient is informed of the lack of scientific evidence and the 'fuzzy logic' that I am obliged to use to help them decide what to do. Most patients in Britain opt for treatment even if this is likely to cause visual loss. As an alternative to serial fundus photography, I have recently started offering biopsy to patients so that those with histological evidence of malignancy would be treated without delay. If genetic studies show lethal abnormality, I would encourage urgent treatment, particularly if the tumour is still small, in case metastatic spread has not started; however, if abnormality of chromosome 3 is not seen in a histologically proven melanoma I would not suggest deferral of treatment in view of the risk of late malignant transformation. At one time, 6p gain was considered to protect against chromosome 3 loss [5], but data from Liverpool show that these two chromosomal abnormalities commonly occur together [2].

If the patient opts to defer treatment, I always make sure that the information provided to the patient is well documented. This is because I believe that it is as important to have written evidence for informed consent for nontreatment as it is to record consent for treatment.

Thin, Equatorial, Choroidal Melanoma

My first choice of treatment for equatorial choroidal melanomas is brachytherapy, because it is less invasive than surgical resection, more reliable than phototherapy and less damaging to superficial tissues than proton beam radiotherapy.

With regard to the type of applicator, I prefer ruthenium to iodine plaques. There are several reasons for this. First, the beta irradiation delivered by ruthenium has a relatively limited range thereby minimising any risk of collateral damage to healthy ocular tissues. Furthermore, this treatment is convenient for patients who come from afar because the applicator can be used immediately. This means that most patients are able to have the plaque inserted the day after their initial assessment at my unit so that they do not need to return home to wait for the plaque to be prepared. With the iodine plaque, patients need to wait several days for the implant to be loaded with the radioactive seeds. The ruthenium plaque has a shelf life of about a year so that it can be re-used many times, making it relatively inexpensive. I also find the ruthenium plaque easier to handle than the iodine applicator, which is more bulky. However, the ruthenium plaque has several disadvantages. The radioactive surface is very delicate so that if it is scratched the plaque can become unusable, which causes considerable disruption of the oncology service and inconvenience to patients until new plaques can be ordered. Secondly, it is not possible to modify the applicator so that in some patients a large area of normal choroid and retina may be irradiated, unlike iodine applicators, which allow better tailoring of the dosimetry to the tumour, simply by adjusting the number and distribution of the radioactive seeds. In practice, the fixed area of irradiation does not seem to affect outcome adversely, especially if the plaque is positioned eccentrically so that it is the anterior choroid that is irradiated. The wide safety margin that results serendipitously reduces the risk of tumour recurrence from diffuse and subclinical anterior tumour spread. Ruthenium plaque radiotherapy is advocated only for tumours up to 5 mm in height. Others have reported success with thicker tumours, despite the fact that the apical part of the tumour does not receive a tumouricidal dose of radiation [6]. It is believed that such thick tumours respond to the ruthenium plaque radiotherapy because the very high basal dose of radiation obliterates the blood supply to the tumour, inducing tumour necrosis. Iodine plaques deliver higher doses to the apical zones of thick tumours, albeit with a greater risk of causing collateral damage to healthy ocular tissues. There is scope for studies comparing ruthenium with iodine plaques.

In some centres, small, equatorial tumours are treated with strontium or palladium applicators [7, 8]. Some authors would treat these with proton beam radiotherapy, which they consider to be superior to brachytherapy in all patients [9].

It is tempting to leave some patients without treatment if their tumour is asymptomatic, even if the clinical features strongly suggest malignancy. I tried this with a few patients at a time when I still believed that ocular treatment never influences survival; however, most tumours grew, making treatment necessary so that I have become less enthusiastic about this non-interventionist policy.

Small, Posterior Choroidal Melanoma

If I am not confident that I can reliably treat a tumour with brachytherapy without causing collateral damage to the optic nerve or fovea, my choice of treatment is proton beam radiotherapy. This happens when the tumour extends close to both the optic disc and fovea, especially if it has an awkward shape. Such an approach is one of the main reasons why my incidence of local tumour recurrence after brachytherapy is only 4% [10]. Occasionally, I obtain the patient's permission to defer the choice of treatment until the time of surgery, deciding between brachytherapy and proton beam radiotherapy only once I have inserted the plaque template and checked its position in relation to the tumour, optic disc and fovea. With proton beam radiotherapy, it is not necessary to suture the tantalum markers precisely at the tumour margins, as long as the marker-marker, marker-tumour and marker-limbus measurements are accurate. Such marker insertion is therefore easier than posterior plaque placement and with computer modelling it is possible to identify any inaccuracies and adjust the dosimetry accordingly. Another advantage of proton beam radiotherapy is that the beam can be collimated according to the shape of the tumour. With tumours extending to within a disc diameter of the optic disc edge, the recommended 2.5-mm safety margin results in a high dose of radiation being delivered to the optic nerve so that optic neuropathy almost inevitably occurs. In some patients, I therefore reduce the lateral and distal safety margins, as long as the tumour is not diffuse and provided that during simulation the patient can keep the eye still. This is done only if the patient gives informed consent, accepting the increased chance of local tumour recurrence and the persistent albeit reduced risk of optic neuropathy. There is scope for further investigation into proton beam safety margins and how these might be adjusted with tumour size.

If the tumour extends close or up to the disc margin, so that any radiotherapy would inevitably cause optic neuropathy, then I would consider endoresection, but only if the patient is highly motivated to conserve vision and accepts the controversial nature of this procedure. The chances of conserving central visual acuity diminish if the tumour extends temporal to the vertical meridians of the optic disc. This procedure is also less likely to be successful if the tumour has a basal dimension exceeding 10–11 mm or if it has diffuse margins. Retinal perforation by the tumour is not a contraindication, especially as retinotomy is routinely performed to gain access to the tumour. Some authors have advocated a retinal flap but my impression is that such a modification has little benefit unless retinal translocation is performed. This may be useful if the tumour extends to the subfoveal choroid and if the vision in the fellow eye is poor. The chances of conserving central vision are good if the tumour does not extend close to the fovea. The most common complications are those of the vitrectomy and are the same as those of other vitreoretinal procedures (e.g. entry site tears). Proliferative vitreoretinopathy does not occur in the absence of rhegmatogenous retinal detachment. As with other forms of conservative therapy, marginal tumour recurrence can arise from invisible tumour remnants in the adjacent uvea. I therefore perform adjunctive brachytherapy a few weeks after the endoresection if histology shows high-grade malignancy. A major concern is that of tumour dissemination around the vitreous cavity, but my impression is that this occurs only if local tumour recurrence occurs in the sclera or uvea and if the treatment of such recurrent tumour is delayed [11]. As with transpupillary thermotherapy, retrobulbar tumour extension can occur from intrascleral tumour deposits, but this is rare [12]. Episcleral, subconjunctival tumour seeding can occur, but with an incidence of less than 1%. Some authors have advocated neo-adjuvant radiotherapy because they are concerned about seeding and metastasis. The rarity of local recurrence after endoresection suggests that few patients benefit from such radiotherapy, which can cause significant morbidity. There is scope for randomised trials comparing endoresection with and without such radiotherapy, but such studies would be difficult to perform because of the rarity of endoresection.

On hearing about the complications of endoresection and of radiotherapy, some patients with juxtapapillary melanoma opt for enucleation, which is performed in the usual manner.

Large, Choroidal Melanoma

The choice of treatment for tumours exceeding 5 mm in thickness is complex. My preference is proton beam radiotherapy, which is less invasive than transscleral local resection, more reliable than ruthenium plaque radiotherapy and less likely than iodine plaque brachytherapy to damage the optic nerve and fovea (unless of course the tumour extends close to these structures). Transscleral local resection is more likely to be selected if there is a high risk of severe, exudative, retinal detachment and neovascular glaucoma, as indicated by a large tumour size and/or extensive retinal detachment. Local resection is also preferred if proton beam radiotherapy is likely to cause canalicular damage and epiphora because the tumour is located inferonasally. Conversely, I tend to select proton beam radiotherapy if resection is expected to be difficult because the tumour has a narrow base, or if it involves more than 1 clock hour of the ciliary body or if it has a bulky, posterior, extra-ocular extension. Other contraindications to surgical resection include: poor general health precluding hypotensive anaesthesia, diffuse melanoma and optic disc involvement. If patients with any of these conditions are also considered unsuitable for proton beam radiotherapy, they are treated by enucleation. Other centres treat large choroidal melanomas with various forms of brachytherapy or with stereotactic radiotherapy [13].

Ciliary Body Melanoma

With ciliary body tumours, treatment selection is based mostly on the extent of involvement of the ciliary body, iris and angle as well as on the tumour thickness. Transscleral resection has the advantage of providing tissue for diagnosis and prognostic studies and is therefore my preferred option [14]. However, local resection is likely to cause excessive morbidity if the circumferential tumour spread exceeds 2 clock hours or if the extent of iris involvement would require excision of the iris sphincter. Surgical resection is difficult if the tumour is thin and is more likely to be complicated by local recurrence if the tumour margins are diffuse. These features are therefore indications for some form of radiotherapy. Proton beam radiotherapy provides a uniform dose of radiation to the tumour and allows modulation of the beam so as to minimise radiation of the cornea and lens. Brachytherapy may be more convenient for patients living far from Liverpool, but requires two surgical procedures and delivers relatively high doses of radiation to the cornea and sclera, thereby increasing the chances of keratopathy and scleral necrosis, particularly with thick tumours.

Iris Melanoma

Iridectomy can cause troublesome photophobia and a significant cosmetic defect. Although these are treatable with a cosmetic contact lens, pupilloplasty or an artificial iris implant, I prefer proton beam radiotherapy [15]. This avoids the problems of an iris coloboma and is simple to administer. Proton beam radiotherapy is also possible with extensive iris tumours and good results can follow treatment of the entire anterior segment. Cataract is to be expected after such treatment and glaucoma can also develop, possibly requiring a drainage procedure, which should be safe if the tumour is sterile [16]. Some patients develop limbal stem cell deficiency. Some authors therefore harvest limbal stem cells before starting the radiotherapy, replacing these cells once the treatment has been completed. I prefer to adopt a wait-and-see approach, with the possibility of performing an autograft of limbal stem cells from the fellow eye, although so far this has not been necessary. Patients are informed that experience with proton beam radiotherapy of extensive iris melanomas is limited so that long-term outcome is uncertain. Others have reported on brachytherapy of iris melanomas using customised plaques [17].

Method of Treatment Selection

There is no consensus about how patients should be counselled about the various therapeutic options. Some ocular oncologists follow a 'paternalistic' approach, recommending their preferred treatment to the patient, even if outcome data on rival techniques are lacking. This method is believed to instil confidence and reassurance in patients who are said to expect such decisive advice from the specialist. I prefer a more consensual way of selecting treatment, exploring the patients' needs, wishes and fears and sharing with them any uncertainties about which rival treatment is most likely to be successful. I do not wish to imply that my methods are superior in any way. This important aspect of care merits further investigation not only with regard to treatment but also with respect to prognostication [18].

I will briefly describe my technique for treatment selection. The patient is seated before me, with my desk at my right elbow. Any relatives are invited to sit next to the patient. I first describe the eye and the tumour, reporting the tumour size and location, and this is done with the aid of a plastic model of the eye and with photographs, which are displayed on a large computer screen on the desk. Next, I explain what might happen if the tumour is left untreated and define what I consider to be the objectives of treatment. At this point, I explain that the survival probability depends on the tumour's genetic abnormalities, emphasising that a significant proportion of tumours have already metastasised by the time they are diagnosed and treated. Patients are informed that the type of ocular treatment is unlikely to influence life expectancy if the tumour has already spread. I emphasise that enucleation is therefore unlikely to improve their survival prospects as compared to conservative forms of treatment. I also mention that ocular treatment may prevent the onset of metastatic spread if such spread has not yet occurred. It is explained that an unknown proportion of uveal melanomas never develop metastatic capability, even if left untreated. I then proceed to describe what I consider to be the best treatment for the patient I am speaking to, discussing the logistics and the likely outcome. If there is equipoise between two or more forms of treatment, I discuss the advantages and disadvantages of each treatment. I then ask patients about their priorities and concerns and the best treatment for the individual concerned usually becomes clear. When patients cannot make up their mind, I ask them probing questions, such as how they might feel if a particular complication occurred, how much they mind travelling to and from the oncology centre for treatment and follow-up assessments, how they would feel if the eye ever needed to be removed, and so on. The answers to these questions usually reveal the treatment that is most suitable for that patient. Rarely, the patient is unwilling or unable to decide what treatment to have and I then recommend a treatment. Some patients request more time to think about matters and this time is granted, of course, but only after asking for the patient's tentative decision about treatment, emphasising that there is always the freedom to change one's mind afterwards. A checklist including all major outcomes and complications is then reviewed, with estimates of the 10-year probability of each outcome being written on the form and discussed with the patient. The entire conversation is recorded using a digital or tape recorder, according to the patient's preference so that a CD-ROM or audio cassette recording is given to the patient. Audits have shown that this practice is very popular with patients, helping them to remember what is said and enabling them to share information with relatives and friends who could not be present at their consultation [19]. Some patients have even passed on their recording to their family doctor or referring ophthalmologist, thereby empowering their practitioner to provide them with better care and advice. I suspect that these recordings may even have prevented misunderstandings that might have resulted in unfounded dissatisfaction or complaint. Immediately after my consultation, the patient is escorted to an adjacent room by a nurse, who ensures that the patient has a good understanding of the situation, confirms that all the patient's needs are met, and provides information leaflets relevant to the individual's condition and treatment. Finally, all patients receive copies of the report sent to the referring ophthalmologist and family doctor, with this document stating the selected treatment and the reasons for such selection. After treatment, all patients are invited to complete a quality of life questionnaire on the anniversary of their treatment so that in time it will be possible to understand how patients feel after various forms of ocular treatment according to the outcomes that are achieved. Such information should in future inform decision-making when selecting treatment.

Summary and Conclusions

In summary, when selecting treatment for uveal melanoma, it is first necessary to decide between treatment and non-treatment. Next, one needs to choose between enucleation and ocular conservation. If treatment is to be conservative, then it is necessary to select between radiotherapy, resection and phototherapy and various combinations of these modalities. Treatment needs to be tailored to the size and location of the tumour, the secondary effects of this tumour on the rest of the eye, the status of the fellow eye and the patient's needs, wishes and fears.

Although it is desirable for treatment selection to be evidence based, the required evidence is either inadequate or completely lacking. This is because of the logistical difficulties preventing randomised studies of treatment versus non-treatment, enucleation versus conservative therapy, and of one type of conservative treatment versus another. The Collaborative Ocular Melanoma Study is widely regarded as confirming that iodine plaque radiotherapy is as effective as enucleation in prolonging life; however, the sample size, although impressive, was insufficient because many patients already had metastases at the time of treatment (as evidenced by their short survival times) and many others had a non-lethal tumour (as indicated by recent studies on genetic typing of uveal melanoma) [20]. Future studies investigating the impact of ocular treatment on survival will need to include only patients whose prognosis can be estimated according to clinical stage, histological grade and genetic type. Another limitation of the published literature is the lack of standardised methods for defining tumour dimensions and extent and for measuring outcomes. There is also much variation in the way that the different treatments are undertaken. As for quality of life studies, these have tended to merge all patients into one cohort whereas it is likely that different age groups and sexes will each respond in a particular manner. Furthermore, quality of life is determined not only by the type of treatment that is administered but also whether or not such treatment is successful in conserving vision and preventing local recurrence. To my knowledge, patient-centred outcomes have not been reported according to these factors.

For these reasons, in many patients, the selection of treatment for uveal melanoma is necessarily based on inadequate data. In other words, the choice of treatment is to some extent a gamble rather than the rational decision one would like to have. Yet, there are few guidelines for coping with uncertainty when planning patient care. There is scope for comparing different methods of treatment selection and assessing these in terms of patient satisfaction with this process.

From what I have heard from my patients, I have gained the impression that there is a tendency for practitioners to mention only the therapeutic modalities that are available in their clinic or hospital, without discussing forms of treatment that are available only at distant locations. For example, I recently saw a patient who had been offered plaque radiotherapy at another centre but who asked to be referred to Liverpool only because she wanted treatment without delay (there being staff sickness at the other centre). In Liverpool, she was dismayed to discover that proton beam radiotherapy had not been mentioned to her at the previous hospital so that the consent she had given for treatment was not as informed as she would have liked. Some might argue that there is no point in mentioning treatments that are not available locally as this might upset patients who cannot afford to travel abroad for such therapy; however, such an egalitarian approach does not give due choice to those who have the means to obtain specialised treatment elsewhere. This is a controversial subject that merits further study so that useful guidelines and recommendations can be prepared.

In conclusion, treatment selection for uveal melanoma is difficult, not only because of the wide variety of clinical scenarios and the range of therapeutic options but also because there is so little guidance on the decision-making process, particularly when scientific evidence is lacking. In this chapter, I have discussed my personal approach to treatment selection and identified scope for improvement in this important aspect of patient care.

References

1 Damato B: Does ocular treatment of uveal melanoma influence survival? Br J Cancer 2010;103:285–290.
2 Damato BE, Dopierala JA, Coupland SE: Genotypic profiling of 452 choroidal melanomas with multiplex ligation-dependent probe amplification. Clin Cancer Res 2010;16:6083–6092.
3 Callejo S, Coupland SE, Damato B: Sudden growth of a choroidal melanoma and multiplex ligation-dependent probe amplification findings suggesting late transformation to monosomy 3 type. Arch Ophthalmol 2011;129:958–960.
4 Damato B, Coupland SE: A reappraisal of the significance of largest basal diameter of posterior uveal melanoma. Eye (Lond) 2009;23:2152–2160.
5 Parrella P, Sidransky D, Merbs SL: Allelotype of posterior uveal melanoma: implications for a bifurcated tumor progression pathway. Cancer Res 1999;59: 3032–3037.
6 Kaiserman N, Kaiserman I, Hendler K, Frenkel S, Pe'er J: Ruthenium-106 plaque brachytherapy for thick posterior uveal melanomas. Br J Ophthalmol 2009;93:1167–1171.
7 Van Ginderdeuren R, Van Limbergen E, Spileers W: 18 years' experience with high dose rate strontium-90 brachytherapy of small to medium sized posterior uveal melanoma. Br J Ophthalmol 2005;89: 1306–1310.
8 Finger PT, Chin KJ, Duvall G: Palladium-103 ophthalmic plaque radiation therapy for choroidal melanoma: 400 treated patients. Ophthalmology 2009; 116(4):790–796.
9 Gragoudas ES: Proton beam irradiation of uveal melanomas: the first 30 years. The Weisenfeld Lecture. Invest Ophthalmol Vis Sci 2006;47:4666–4673.
10 Damato B, Patel I, Campbell IR, Mayles HM, Errington RD: Local tumor control after (106)Ru brachytherapy of choroidal melanoma. Int J Radiat Oncol Biol Phys 2005;63:385–391.
11 Hadden PW, Hiscott PS, Damato BE: Histopathology of eyes enucleated after endoresection of choroidal melanoma. Ophthalmology 2004;111:154–160.
12 Damato B, Wong D, Green FD, Mackenzie JM: Intrascleral recurrence of uveal melanoma after transretinal 'endoresection'. Br J Ophthalmol 2001; 85:114–115.
13 Dunavoelgyi R, Dieckmann K, Gleiss A, et al: Local tumor control, visual acuity, and survival after hypofractionated stereotactic photon radiotherapy of choroidal melanoma in 212 patients treated between 1997 and 2007. Int J Radiat Oncol Biol Phys 2010, E-pub ahead of print.
14 Damato EM, Damato B, Sibbring JS, Coupland SE: Ciliary body melanoma with partial deletion of chromosome 3 detected with multiplex ligation-dependent probe amplification. Graefes Arch Clin Exp Ophthalmol 2008;246:1637–1640.
15 Damato B, Kacperek A, Chopra M, Sheen MA, Campbell IR, Errington RD: Proton beam radiotherapy of iris melanoma. Int J Radiat Oncol Biol Phys 2005;63:109–115.

16 Tan AN, Hoevenaars JG, Webers CA, Damato B, Beckers HJ: Baerveldt implant for secondary glaucoma due to iris melanoma. Clin Ophthalmol 2010; 4:407–409.
17 Shields CL, Naseripour M, Shields JA, Freire J, Cater J: Custom-designed plaque radiotherapy for nonresectable iris melanoma in 38 patients: tumor control and ocular complications. Am J Ophthalmol 2003; 135:648–656.
18 Cook SA, Damato B, Marshall E, Salmon P: Reconciling the principle of patient autonomy with the practice of informed consent: decision-making about prognostication in uveal melanoma. Health Expect 2010, E-pub ahead of print.
19 Ah-Fat FG, Sharma MC, Damato BE: Taping outpatient consultations: a survey of attitudes and responses of adult patients with ocular malignancy. Eye 1998;12:789–791.
20 Damato B: Legacy of the collaborative ocular melanoma study. Arch Ophthalmol 2007;125:966–968.

Bertil E. Damato, MD, PhD, FRCOphth
Ocular Oncology Service, Royal Liverpool University Hospital
Prescot Street
Liverpool L7 8XP (UK)
Tel. +44 151 706 3973, E-Mail bertil@damato.co.uk

Jager MJ, Desjardins L, Kivelä T, Damato BE (eds): Current Concepts in Uveal Melanoma.
Dev Ophthalmol. Basel, Karger, 2012, vol 49, pp 27–40

Ruthenium-106 Brachytherapy

Jacob Pe'er

Department of Ophthalmology, Jonas Friedenwald Professor of Ophthalmic Research, Hadassah-Hebrew University Medical Center, Jerusalem, Israel

Abstract

Brachytherapy is the most common method for treating uveal melanoma, and currently the ruthenium-106 (Ru-106) and iodine-125 (I-125) applicators are the most frequently used. Ru-106 applicators were introduced by Prof. Peter Lommatzsch in the 1960s, and since then have been used widely by many ocular oncologists, mainly in Europe. The Ru-106 isotope is a beta ray (electron) emitter, and as such it has a limited depth of penetration. This is the reason why many experts use Ru-106 applicators for tumors with a maximal thickness of up to 7.0 mm, although others use it successfully for thicker tumors. The Ru-106 applicators are manufactured commercially and have a half-life of about 1 year. Ru-106 brachytherapy for uveal melanoma provides excellent local control rates and eye preservation with a relatively low recurrence rate. The main advantage of Ru-106 over other isotopes is the better preservation of vision in the treated eye, and less damage to the healthy parts of the eye due to its limited range of radiation. This can also be achieved by positioning the Ru-106 plaque eccentrically, away from the macula and optic nerve head. Ru-106 brachytherapy can be used in combination with other methods of treatment of uveal melanoma, such as local resection or transpupillary thermotherapy, and is sporadically combined with other isotopes, such as gamma-emitting cobalt-60 and I-125.

Although alternative methods of treatment have existed for the past 80 years, enucleation was the most common way of treating uveal melanoma until the late 1970s. Over the last 3 decades, enucleation has largely been replaced by 'conservative' methods, aiming to destroy the primary uveal melanoma while conserving the eyeball and as much useful vision as possible.

Among the conservative methods, brachytherapy is the one most commonly used. Brachytherapy (the Greek term 'brachy' means 'short distance') is local irradiation, using implanted radioactive material either within or close to the treated tumor. In 1930, Foster Moore [1] initiated brachytherapy as a modality for treating ocular tumors by intraocular insertion of radon seeds. Stallard [2] followed him, using radon for the treatment of retinoblastoma, and, in 1939, introduced cobalt-60

(Co-60) applicators for the treatment of choroidal melanoma [3]. For 4 decades, Co-60 was the main nuclide used for successfully treating intraocular tumors. However, the high energy of Co-60 caused unwanted biological effects, adversely affecting the healthy tissues of the treated eye and exposing the medical personnel to irradiation. Another isotope was needed, and indeed ruthenium-106 (Ru-106) [4], iodine-125 (I-125) [5], strontium-90 [6], palladium-103 [7], gold-198 [8] and iridium-192 [9] were introduced to ocular oncology. Currently, Ru-106 and I-125 are by far the most commonly used applicators.

Prof. Peter Lommatzsch introduced the application of Ru-106 plaques in East Germany, and started using them for the treatment of uveal melanoma in 1964 [4]. He first used Ru-106 plaques for the treatment of choroidal melanoma [10, 11] and then for ciliary body melanoma [12]. Prof. Lommatzsch investigated the morphological and functional changes of the Ru-106 radiation on rabbit eyes [13] and subsequently the effect of this radiation on various anatomical parts of the treated human eyes [14–17]. He summarized the results of his long experience in a series of papers [18–21]. Many ocular oncologists, mainly in Europe, followed his path and have used the Ru-106 applicators routinely.

The Ru-106 Applicator

The Ru-106 isotope is a beta ray (electron) emitter with a half-life of 373.59 days [22–28]. Gamma radiation contributes only 1% to the total dose of radiation. Ru-106 decays to rhodium-106 while emitting beta rays with a maximum energy of 39 keV. Rhodium-106, with a half-life of 36 s, decays to palladium-106 with a maximum energy of 3.54 MeV and mean energy of 1.5 MeV, providing the effective therapeutic irradiation.

In manufacturing the Ru-106 applicators, the radionuclide is homogeniously deposited on a thin silver foil by electrolysis. This radioactive target is then placed on a 0.9-mm-thick silver plate and sealed by a 0.1-mm silver sheet, facing the direction of the eyeball. This cover has almost no absorption effect on the beta ray emission, while the silver on the convex side absorbs more than 95% of the radioactivity. The diameter of the 16 different types of applicators varies from 11 to 25.4 mm, with a spherical radius of 12–14 mm (fig. 1). The nominal activity of the Ru-106 applicators ranges between 4.1 and 40.3 MBq, equal to 0.11–1.09 mCi, depending on the size and shape of the applicator.

Most uveal melanomas are treated by round applicators with various diameters. Notched plaques have been designed for treating tumors next to the optic nerve head, while crescent-shaped plaques have been designed for anterior tumors, involving the ciliary body and/or iris periphery, in order to prevent irradiation of the cornea.

The Ru-106 applicators have an inactive edge at the peripheral margin, which is reported to be about 0.7–0.8 mm. The dose rate falls significantly near the periphery,

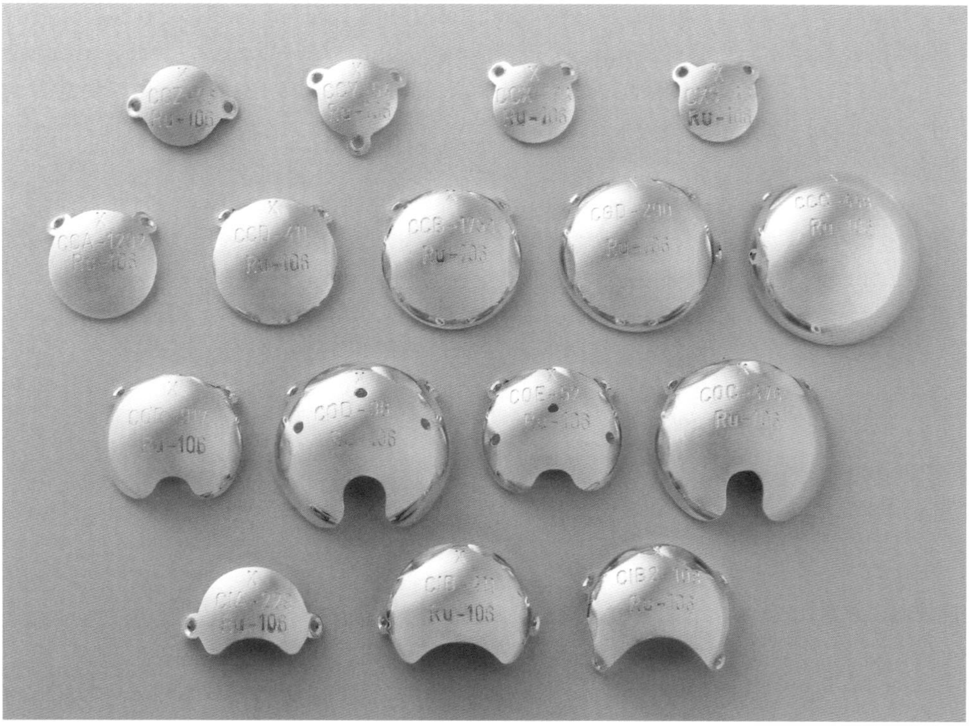

Fig. 1. 16 types of Ru-106 eye applicators (Source: Eckert & Ziegler BEBIG GmbH).

and within about 1 mm of the applicator edge the dose is about 60–70% of the center value.

While being manufactured by Isocommerz (Berlin, Germany) using the ASMW specification (ASMW is the standardization office of the former German Democratic Republic), the relative distribution of activity on the plaque surface was measured only on a few points, and the dose measurements were very inaccurate. The uncertainty, as indicated by the manufacturer's specifications, was ±30%. Since the 1990s, the Ru-106 applicators have been produced by Eckert & Ziegler BEBIG GmbH (Berlin, Germany) and new developments in dosimetry have been realized. From 2002 on the calibration of the Ru-106 applicators has been based on the NIST (National Institute of Standards and Technology, USA) primary standard for absorbed dose to water for beta radiation sources. The current expanded measurement uncertainty of the absolute values is ±20% (95% confidence level) with consideration of a 15% contribution from the NIST calibration (BEBIG).

The dose rate of Ru-106 at the time of calibration by the manufacturer is about 110 mGy/min or about 660 cGy/h at a depth of 1 mm, depending on the type of applicator. Because of the low penetration of electrons, there is a rapid dose falloff. At a depth of 3 mm, the dose rate is about 60 mGy/min, about 30 mGy/min at a depth of 5 mm, and about 10 mGy/min at a depth of 10 mm, which is less than 10% of the dose rate at a depth of 1 mm. That means that in order to give the routinely planned dose of 80–100 Gy to the apex of the tumor in case of a tumor with a height of 8 mm, the sclera and the tumor base will receive more than 10 times more irradiation. Since the half-life of the Ru-106 is about 1 year, after 52 weeks, the dose rate is reduce to about 50%. This means that in order to achieve the planned dose, the treatment time should be double the time necessary when using a new applicator. According to the manufacturer's instructions, the maximal useful life of the applicators is 18 months, and the number of possible sterilization cycles is restricted to 50.

The low range of the electrons emitted by the Ru-106 applicators limits the depth of their penetration into the treated tissue. On the other hand, it has the advantage of sparing the surrounding healthy tissue inside and around the eye, and is important for radiation protection of the medical personnel who handle the applicator, and only minor safety precautions have to be undertaken. Reprocessing prior to usage can be easily performed in an autoclave using the dedicated sterilization container.

The Use of Ru-106 Applicators for Uveal Melanoma

Indications and Doses

Brachytherapy is the most common conservative method of treating uveal, mainly posterior uveal, melanoma. Technically, radioactive applicators can be applied to tumors in almost all locations in the eye, except for tumors that grow over or into the optic nerve. However, tumor size limits the use of these applicators, and when tumors are 'extra large', with a diameter of 20 mm or larger and/or a thickness of 10 mm or more, most experts will recommend enucleation of the tumor-harboring eye.

I-125 applicators with deeper radiation penetration have been used by many, including the COMS group, for tumors up to 10 mm in height [29]. Ru-106 applicators, with their more limited depth of penetration, are usually used for tumors with a maximal thickness of up to 7.0 mm [30, 31], in order to achieve 100 Gy to the tumor apex, without exceeding 1,000 Gy to the tumor base. Some experts calculate a minimal dose of 300–400 Gy to the tumor base, even if the tumor apex will receive more than 100 Gy [31], while for thick tumors, others calculate 1,000 Gy or even up to 1,500 Gy to the base, even if the tumor apex will receive less than what is considered to be a tumoricidal dose (80–100 Gy) [31, 32].

Surgical Techniques

Accurate measurement of the tumor dimensions is very important for selection of the appropriate size and shape of the applicator and for calculation of the irradiation dose and delivery time. The tumor basal diameter is measured by funduscopy and fundus photography, preferably by using a wide-angle camera, and by B-scan ultrasonography. The tumor thickness is measured by A-scan, or, if not possible, by B-scan and high-frequency ultrasonography, or a combination thereof.

The surgery for positioning the plaque over the treated tumor can be performed under either local or general anesthesia. A 360° or partial peritomy is used to expose the sclera in the tumor area. If the tumor is located under a rectus muscle insertion, the muscle should be disinserted after placing sutures to the muscle tendon and measuring the knot-to-limbus distance. Some surgeons disinsert a rectus muscle even if its insertion is not over the tumor base, in order to facilitate their approach to the tumor, mainly when the tumor is located posteriorly. When the tumor is in the posterior pole, the inferior oblique muscle has to be disinserted in order to enable good positioning of the applicator.

The tumor margins are identified by transpupillary or transscleral transillumination, or by indirect ophthalmoscopy with indentation, and are marked by a pen or with an electrical photocoagulator. In many centers, the planning of the Ru-106 plaque positioning is done with the help of a metal nonirradiating plaque, identical in size and shape to the irradiating plaque, or by a plastic template that enables transillumination through a hole in the template for locating the plaque [31]. Several experts use intraoperative ultrasound to confirm the positioning of the template or the plaque [33] and others use postoperative ultrasound, taking into consideration that the eye movements, muscles and optic nerve can push the plaque slightly [34]. When the tumor is close to the fovea or optic nerve, in trying to protect vision, Ru-106 applicators can be positioned eccentrically in relation to the tumor, aligning the posterior edge of the plaque with the posterior tumor margin, relying on the side scatter radiation to treat the tumor margins; this also results in good tumor control [31, 35; Pe'er, unpubl. data]. After positioning the plaque, rectus muscles that had been disinserted are repositioned and the conjunctiva is closed. If an inferior oblique muscle was disinserted, it is usually left unattached.

When the calculated dose of irradiation has been delivered, the Ru-106 plaque is removed in a second surgery, usually after 2–10 days. The patient is invited for clinical examination after a couple of weeks, but the response of the tumor to irradiation can be seen and evaluated only after 2–3 months. Life-long follow-up is needed, and in most centers the recommendation is for patient evaluation, either in the ocular oncology center or by the patient's ophthalmologist, every 6 months.

Results of Ru-106 Brachytherapy

Local Tumor Control

Ru-106 brachytherapy for posterior uveal melanoma provides excellent rates of local control and eye preservation. Over 96% of the treated tumors respond by a decrease in tumor height; the initial rate of the decrease is approximately 3% per month [36]. Most of the tumors do not regress entirely; rather, after 18–24 months, their height stabilizes on an average value of about 60% of the initial height according to one study [36], and on about 70% after 36 months according to another study [37]. Large tumors have a faster initial decrease in height and stabilize at a lower percentage of their initial height as compared to small tumors. When following up the tumors by A-scan ultrasonography, the internal reflectivity of the uveal melanoma increases significantly after Ru-106 brachytherapy from a mean of 30–40% before therapy to 60–70% after 2 years [36, 37]. Larger tumors which initially had a lower internal reflectivity show a slower increase in internal reflectivity compared to smaller tumors.

Although most uveal melanomas treated with Ru-106 brachytherapy exhibit a significant reduction in size, histological evaluation of such irradiated tumors shows, in addition to necrosis and scarring, that most of these tumors harbor viable tumor cells [19, 38–40]. Using immunohistochemical markers such as PC-10 for PCNA and MIB-1 for Ki-67 cells showed proliferating activity [40–42]. However, this activity was significantly lower in tumors that responded well to brachytherapy and did not regrow, compared to tumors that regrew or were not treated by irradiation at all [40].

Tumor Recurrence

The tumor recurrence rate following Ru-106 brachytherapy varies significantly among ocular oncology centers. The recurrence rate was as low as 3–4% in some studies [31, 43] and as high as 11% [44, 45] and 16% [46] in other studies. Tumor recurrence can happen as early as a year and as late as 10 years after a good response. The main risk factor for tumor recurrence is tumor size, its largest diameter and tumor thickness, although tumor location, mainly posterior location, is also of importance. The difference in tumor recurrence can be explained, among other factors, by tumor selection. While in many centers Ru-106 brachytherapy is limited to tumors with a maximal thickness of 5–7 mm, other centers may treat thicker tumors, instead of enucleating them up front [31, 32] (fig. 2). Tumor recurrence was not found to be associated with radiation dose or radiation rate to the tumor base or apex.

Whenever possible, tumor recurrence will be treated by additional brachytherapy, but in those cases with significant recurrence, mainly when the tumor shows a diffuse pattern or when the primary dose of irradiation has already been high, the eye will be enucleated.

Tumor recurrence is one of two main reasons for enucleation after brachytherapy; the other one is ocular complications, as will be discussed later in this chapter. The post-Ru-106 brachytherapy enucleation rate varies among ocular oncology centers, ranging from 4.4% after 5 years [43] to as high as 17–18% [30, 47], similar to the results of the COMS group [48].

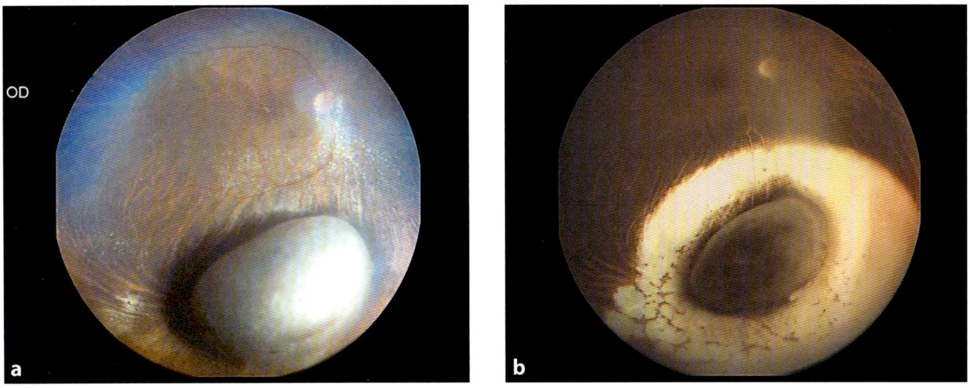

Fig. 2. **a** A large choroidal tumor in the right eye, measuring 9.3 mm in maximal height. **b** The same tumor after Ru-106 brachytherapy, stabilized at a maximal thickness of 4.2 mm and after a follow-up of 5 years.

Visual Outcome

Conservative treatment of uveal melanoma is aimed not only at preserving the eye globe but also, whenever possible, at preserving vision. The loss of visual acuity following brachytherapy depends on several risk factors; the most important ones are reduced initial visual acuity before treatment, and the distance between the posterior tumor margins and the vital ocular structures – the fovea and the optic nerve head [30, 35]. Other important risk factors for losing vision are increased tumor height, older age, and a temporal tumor location [35]. The importance of the total dose to the tumor apex and base and the influence of dose rate on the visual acuity following brachytherapy are controversial [30, 35].

Most centers limit the use of Ru-106 applicators to tumors not higher than 7 mm, but it seems that despite this limitation, the visual acuity results differ among centers. In a follow-up of up to 10 years in one center [49], a visual acuity of 6/21 or better was found in 27%, 6/60 or better in 41% and finger counting in 82% or better of eyes. Another study [30] with a follow-up of 5 years showed that 31% of the patients retained a visual acuity of 6/12 or better, and 49% retained a visual acuity of 6/60 or better. A third study [35] achieved much better results after 9 years of follow-up, in spite of treating tumors of similar size, reaching a visual acuity of 6/12 or better in 55%, 6/60 or better in 57% and finger counting or better in 83%. These good results were probably achieved by this group as well as by others [50] because they positioned the Ru-106 applicators eccentrically in treating tumors that are located close to the macula and optic nerve head (fig. 3). Using eccentrically positioned applicators enables preservation of vision even when treating uveal melanoma thicker than 8 mm by Ru-106 applicators, with 70% of them achieving a visual acuity of 6/60 or better in the eyes that were preserved [32]. When the visual acuity results of Ru-106

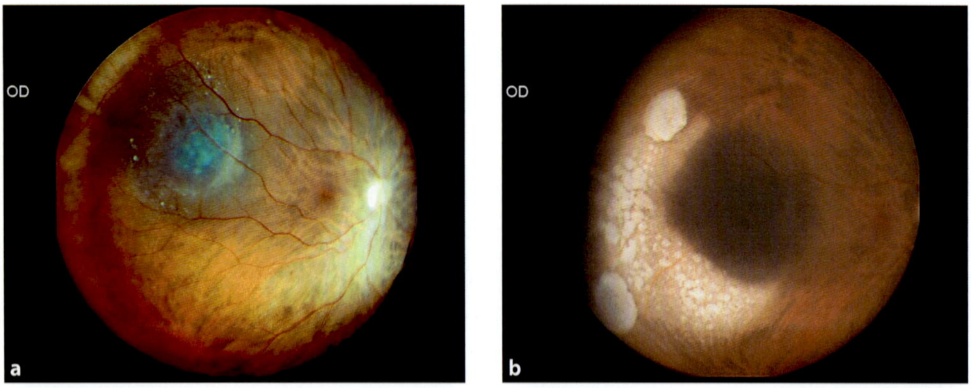

Fig. 3. **a** A choroidal melanoma in the right eye, close to the macular area. **b** The melanoma was treated successfully by positioning of the Ru-106 applicator eccentrically, in order to protect the fovea. The tumor height decreased from 3.1 to 1.1 mm, and the visual acuity remained at 6/6 6.5 years after treatment.

brachytherapy are compared to I-125 brachytherapy, treating tumors of similar size, a better preservation of visual acuity was observed in the Ru-106 patients [26, 35, 44, 51]. This can be explained by the fact that Ru-106 plaques emit beta rays, which have a more limited range than the I-125 gamma source, and thus provide a significantly lower dose to the lens, macula and optic nerve head [26].

Survival

Survival after treatment of uveal melanoma probably does not depend on the method of treatment, but rather on many clinical, histological and genetic risk factors, which are beyond the scope of this chapter. Probably the most important clinical risk factor is tumor size. A meta-analysis of 5 case series of patients treated by Ru-106 brachytherapy [52] showed a 5-year melanoma-related mortality rate of 6% for small and medium tumors and 26% for large tumors. In the same study, 5- and 10-year melanoma-related mortality rates for a balanced set of small, medium and large tumors were 14 and 22%, respectively. Other studies showed an overall 5-year melanoma-related mortality rate of 16 and 14% [47, 53] and a 5-, 10- and 15-year melanoma-related mortality rates of 11.4, 17 and 23%, respectively [45]. These results are somewhat better than the results in the COMS study [54]. Some studies find a relationship between mortality and Ru-106 brachytherapy failure [47], but other studies do not find such a relationship [45, 53]. One study on the impact of dose rate in Ru-106 brachytherapy found significantly less metastases in patients with tumors treated with a higher dose rate (>4 Gy/h) than in those treated with a lower dose rate (<4 Gy/h) [26].

Complications

As happens after irradiating the eye for treating intraocular tumors by other sources, complications also occur in some eyes after Ru-106 brachytherapy, although at lower percentages as compared with I-125 brachytherapy and proton beam radiation [44]. The most common complications are cataract, radiation retinopathy and maculopathy, irradiation optic neuropathy, vitreous hemorrhage, and iris neovascularization (rubeosis iridis) leading in most cases to neovascular glaucoma.

Cataract related to Ru-106 brachytherapy is the most common complication, and in one study its 2-, 3- and 5-year probability was 21, 27 and 37%, respectively [55]. The development of cataract is related to the tumor location in the anterior part of the choroid and ciliary body, when the lens is affected by direct irradiation, and to the tumor size since larger tumors receive higher doses of radiation. Cataract surgery can be performed to cure the radiation cataract.

Iris neovascularization was detected in 12% of the irradiated eyes after Ru-106 brachytherapy [55]. Since it depends on retinal ischemia, it was found to be related to the tumor size, mainly the largest basal tumor diameter. The extent of the retinal detachment accompanying the tumor, and especially persistent retinal detachment, can also be a risk factor. Most of the eyes with iris neovascularization develop neovascular glaucoma, and according to one study, the cumulative 2-, 3- and 5-year probability for neovascular glaucoma was 4, 11 and 21%, respectively [55].

Vitreous hemorrhage is diagnosed sometimes at the time of tumor diagnosis, but more frequently may develop after treatment with brachytherapy. It was found to be related to tumor size, mainly its height, and the cumulative 2-, 3- and 5-year probability was 9, 17 and 24%, respectively [55]. Most vitreous hemorrhages following brachytherapy clear spontaneously, but extensive hemorrhage may require vitrectomy. A need for surgical intervention may indicate a poor prognosis for salvaging the eye.

Radiation maculopathy is diagnosed when retinal vascular changes are identified, such as microaneurysms, hemorrhages, microinfarcts and lipid exudation in the macular area. Macular edema is also commonly seen, and ocular coherent tomography enables the diagnosis of edema in its early stages. Diabetic patients are at greater risk for developing radiation maculopathy. Radiation maculopathy may develop several months after brachytherapy, but usually develops 1–2 years after treatment. In one study, the 2- and 3-year probability of developing maculopathy after Ru-106 brachytherapy was 15 and 30%, respectively [55]. A posterior location of the melanoma and its proximity to the fovea are the main risk factors for developing radiation maculopathy. In recent years, radiation maculopathy is often being treated by intravitreal injections of bevacizumab (Avastin) or ranibizumab (Lucentis), as is done in other vascular maculopathies, with good results [56; Pe'er, unpubl. data]. Periocular injections of triamcinolone were shown by one group to be highly effective in preventing postbrachytherapy macular edema [57].

Radiation optic neuropathy is less common than radiation maculopathy, and the 2- and 3-year probability was found in one study to be 10 and 12%, respectively [55]. The proximity of the tumor to the optic disk margin is the main risk factor for the development of optic neuropathy, which rarely develops when the posterior tumor margin is located more than 4 mm from the optic disk.

Scleral melting following Ru-106 brachytherapy is seen almost only in the anterior sclera, over the ciliary body, when using a high dose of irradiation to the sclera. Scleral melting over choroidal melanoma is extremely rare. In the past, eyes with significant scleral melting were enucleated, but currently, since it was found that the most external part of the irradiated tumor is the most necrotic, enucleation is performed less frequently and the eye can usually be conserved. A possible management for scleral melting is scleral graft.

Motility disorders may be complications of brachytherapy, although usually they are temporary and resolve within several months [58]. These disorders are more frequent when a muscle has to be disinserted during surgery for implantation of the radioactive plaque. When the treated eye is a seeing eye, diplopia can result, although it is almost always temporary and no treatment is needed.

The development of complications after Ru-106 brachytherapy is less common than after I-125 brachytherapy and proton beam radiation, probably because of the more limited distance of the beta radiation. One study showed that the mean dose to the macula, optic nerve head and lens was 18, 53 and 89% less, respectively, with Ru-106 compared to I-125 [27]. Complications to the eyelid and ocular surface, as sometimes seen after proton beam radiation, are rarely observed after Ru-106 brachytherapy.

Ru-106 Brachytherapy for Iris Melanoma

Melanoma of the iris alone, or iridociliary melanoma, can be treated very effectively by Ru-106 brachytherapy [59]. In such cases, small round applicators or crescent-shaped applicators are commonly used. Complications after radiation to the iris and ciliary body are usually confined to the anterior segment. Corneal erosions can occur temporarily, and peripheral corneal opacity can result. Cataract is obviously very common, occurring in most of these patients. Irradiation to the anterior segment often causes anterior uveitis, which should be treated in advance by high doses of corticosteroids and cycloplegics, in order to prevent its complications.

Combined Use of Ru-106 Brachytherapy with Other Treatment Modalities

Ru-106 applicators are usually used alone for the treatment of uveal melanoma, but in some centers they have been used in combination with other known modalities of treatment for uveal melanoma.

Combined Ru-106 brachytherapy and transpupillary thermotherapy for treating uveal melanoma is sometimes used, and is termed 'sandwich therapy' [60]. It may have several advantages, as the two methods of treatment are complementary to each other. This combination enables the treatment of thick tumors by treating the apex with transpupillary thermotherapy, reducing the tumor height, and making it possible to use a lower dose of irradiation via the Ru-106 applicator. Transpupillary thermotherapy can be complementary to brachytherapy when there is a suspicion that the tumor margins are not properly covered by the Ru-106 applicator or if recurrence in the tumor periphery is suspected [31].

Ru-106 brachytherapy can be used as a complementary treatment for uveal melanoma after external or internal local tumor resection, in order to eradicate any possible tumor remnants in the sclera which could cause local recurrence [61, 62].

Combined binuclide applicators of beta-emitting Ru-106 and gamma-emitting Co-60 or I-125 are rarely used [63, 64]. The purpose of such a combination is to have the advantage of the relatively higher dose of gamma radiation to the tumor apex with a higher dose of beta radiation to the tumor base, achieving a high dose to both while minimizing collateral damage to the uninvolved ocular tissue.

Conclusions

Ru-106 brachytherapy provides an excellent rate of local control of uveal melanoma and a high rate of eye preservation with a relatively low rate of recurrence. Because of the limited depth of penetration of the beta rays (electrons), many experts use Ru-106 applicators for tumors with a maximal thickness of up to 7.0 mm. However, others use it successfully for thicker tumors. The main advantage of the Ru-106 over other isotopes is the less damage it causes to the healthy parts of the eye and the better preservation of the vision in the treated eye. This can be enhanced by positioning the Ru-106 applicator eccentrically, away from the macula and the optic nerve head. Ru-106 brachytherapy can also be used in combination with other methods of treatment of uveal melanoma.

References

1 Moore RF: Choroidal sarcoma treated by the intraocular insertion of radon seeds. Br J Ophthalmol 1930;14:145–152.

2 Stallard HB: A new technique for the application of radon seeds to the sclera in the treatment of glioma retinae. Br J Ophthalmol 1938;22:604–608.

3 Stallard HB: Radiotherapy for malignant melanoma of the choroid. Br J Ophthalmol 1966;50:147–155.

4 Lommatzsch P, Vollmar R: A new way in the conservative therapy of intraocular tumors by means of beta-irradiation (ruthenium-106) with preservation of vision. Klin Monbl Augenheilkd 1966;148:682–699.

5 Packer S, Rotman M: Radiotherapy of choroidal melanoma with iodine-125. Ophthalmology 1980; 87:582–290.

6 van Ginderdeuren R, van Limbergen E, Spileers W: 18 years' experience with high dose rate strontium-90 brachytherapy of small to medium sized posterior uveal melanoma. Br J Ophthalmol 2005;89: 1306–1310.
7 Finger PT, Buffa A, Mishra S, Berson A, Bosworth JL, Vikram B: Palladium 103 plaque radiotherapy for uveal melanoma. Clinical experience. Ophthalmology 1994;101:256–263.
8 Chaudhry IA, Liu M, Shamsi FA, Arat YO, Shetlar DJ, Boniuk M: Corneoscleral necrosis after episcleral Au-198 brachytherapy of uveal melanoma. Retina 2009;29:73–79.
9 Shields CL, Shields JA, Karlsson U, Markoe AM, Brody LW: Reasons for enucleation after plaque radiotherapy for posterior uveal melanoma. Clinical findings. Ophthalmology 1989;96:919–923.
10 Lommatzsch PK: Experiences in the treatment of malignant melanoma of the choroid with ^{106}Ru-^{106}Rh beta-ray applicators. Trans Ophthalmol Soc UK 1973;93:119–132.
11 Lommatzsch P: Treatment of choroidal melanomas with ^{106}Ru/^{106}Rh beta-ray applicators. Survey Ophthalmol 1974;19:85–100.
12 BallinR, Lommatzsch PK, Drost H, Ratajek B: A beta ray applicator (106Ru/106Rh) in the treatment of ciliary body melanomas. Klin Monbl Augenheilkd 1985;187:144–146.
13 Lommatzsch P: Morphological and functional changes in the rabbit eye on exposure of the dorsal bulb section to beta rays (^{106}Ru-^{106}Rh). Von Graefes Arch Klin Exp Ophthalmol 1968;176:100–125.
14 Lommatzsch P, Klug H: Ultrastructural changes of the retina damaged by radiation. Ber Zusammerkunft Dtsch Ophthalmol Ges 1966;67:158–163.
15 Lommatzsch P: On the problem of radiation cataract in irradiation of the anterior eye segment. Klin Monbl Augenheilkd 1967;150:683–688.
16 Lommatzsch PK, Ballin RE, Helm W: Fluorescein angiography in the follow-up study of choroidal melanoma after ^{100}Ru/^{106}Rh plaque therapy. Retina 1987;7:148–155.
17 Lommatzsch PK, Albert W, Lommatzsch R, Rohrwacher F: Radiation effects on the optic nerve observed after brachytherapy of choroidal melanomas with ^{106}Ru/^{106}Rh plaques. Graefes Arch Clin Exp Ophthalmol 1994;232:482–487.
18 Lommatzsch PK: β-Irradiation of choroidal melanoma with ^{106}Ru-^{106}Rh applications. 16 years' experience. Arch Ophthalmol 1983;101:713–717.
19 Lommatzsch PK: Results after β-irradiation (^{106}Ru-^{106}Rh) of choroidal melanoma with applications: 20 years' experience. Br J Ophthalmol 1986;70:844–851.
20 Lommatzsch PK, Kirsch IH: ^{106}Ru/^{106}Rh plaque radiotherapy for malignant melanomas of the choroid. Doc Ophthalmol 1988;68:225–238.
21 Lommatzsch PK, Werschnik C, Schuster E: Long-term follow-up of Ru-106/Rh-106 brachytherapy for posterior uveal melanoma. Graefes Arch Clin Exp Ophthalmol 2000;238:129–137.
22 Davelaar J, Schaling DF, Hennen LA, Broerse JJ: Dosimetry of ruthenium-106 eye applicators. Med Phys 1992;19:691–694.
23 Menapace R, Binder W, Chiari A: Results and implications of high-resolution surface dosimetry of ruthenium-106 eye applicators. Ophthalmologica 1992;204:93–100.
24 Laube T, Flühs D, Kessler C, en fisica L, Bornfeld N: Determination of surgeon's absorbed dose in iodine 125 and ruthenium 106 ophthalmic plaque surgery. Ophthalmology 2000;107:366–369.
25 Chan MF, Fung AYC, Hu Y-C, Chui C-S, Amols H, Zaider M, Abramson D: The measurement of three dimensional dose distribution of a ruthenium-106 ophthalmological applicator using magnetic resonance imaging of BANG polymer gels. J Appl Clin Med Phys 2001;2:85–89.
26 Mossbock G, Rauscher T, Winkler P, Kapp KS, Langmann G: Impact of dose rate on clinical course in uveal melanoma after brachytherapy with ruthenium-106. Strahlenther Onkol 2007;10:571–575.
27 Wilkinson DA, Kolar M, Fleming PA, Singh AD: Dosimetric comparison of ^{106}Ru and ^{125}I plaques for treatment of shallow (≤5 mm) choroidal melanoma lesions. Br J Radiol 2008;81:784–789.
28 Eckert and Ziegler BEBIG GmbH: Ru-106 Eye Applicators: Instructions for Use (translation of document TD04_000). 2009.
29 The Collaborative Ocular Melanoma Study Group: The COMS randomized trial of iodine-125 brachytherapy for choroidal melanoma. 3. Initial mortality findings. COMS report No 18. Arch Ophthalmol 2001;119:969–982.
30 Bergman L, Nilsson B, Lundell G, Lundell M, Seregard S: Ruthenium brachytherapy for uveal melanoma, 1979–2003: survival and functional outcomes in the Swedish population. Ophthalmology 2005;112:834–840.
31 Damato B, Patel I, Campbell IR, Mayles HM, Errington RD: Local tumor control after Ru-106 brachytherapy of choroidal melanoma. Int J Radiat Oncol Biol Phys 2005;63:385–391.
32 Kaiserman N, Kaiserman I, Hendler K, Frenkel S, Pe'er J: Ruthenium-106 plaque brachytherapy for thick posterior uveal melanomas. Br J Ophthalmol 2009;93:1167–1171.

33 Tanandeh H, Chaudhry NA, Murray TG, Ehlies F, Hughes R, Scott IU, Markoe AM: Intraoperative echographic localization of iodine-125 episcleral plaque for brachytherapy of choroidal melanomas. Am J Ophthalmol 2000;125:199–204.

34 Anteby I, Pe'er J: Need for confirmation of positioning of ruthenium plaques by postoperative B-scan ultrasonography. Ophthalmic Surg Lasers 1996;27: 1024–1029.

35 Damat B, Patel I, Campbell IR, Mayles HM, Errington RD: Visual acuity after ruthenium-106 brachytherapy of choroidal melanoma. Int J Radiat Oncol Biol Phys 2005;63:392–400.

36 Kaiserman I, Anteby I, Chowers I, Blumenthal EZ, Kliers I, Pe'er J: Changes in ultrasound findings in posterior uveal melanoma after ruthenium-106 brachytherapy. Ophthalmology 2002;109:1137–1141.

37 Georgopoulos M, Zehetmayer M, Ruhswurm I, Toma-Bstaendig S, Segur-Eltz N, Sacu S, Menapace R: Tumour regression of uveal melanoma after ruthenium-106 brachytherapy or stereotactic radiotherapy with gamma knife or linear accelerator. Ophthalmologica 2003;217:315–319.

38 Klaus H, Lommatzsch PK, Fuchs U: Histopathology studies in human malignant melanomas of the choroid after unsuccessful treatment with $^{106}Ru/^{106}Rh$ ophthalmic applicators. Graefes Arch Clin Exp Ophthalmol 1991;229:480–486.

39 Messmer E, Bornfeld N, Foerster M, Schilling H, Wessing A: Histopathologic findings in eyes treated with a ruthenium plaque for uveal melanoma. Graefes Arch Clin Exp Ophthalmol 1992;230:391–396.

40 Pe'er J, Stefani FH, Seregard S, Kivela T, Lommatzsch P, Prause JU, Sobottka B, Damato B, Chowers I: Cell proliferation activity in posterior uveal melanoma after Ru-106 brachytherapy: an EORTC ocular oncology group study. Br J Ophthalmol 2001;85: 1208–1212.

41 Schilling H, Sehu KW, Lee RW: A histologic study (including DNA quantification and Ki-67 labeling index) in uveal melanoma after brachytherapy with ruthenium plaques. Invest Ophthalmol Vis Sci 1997; 38:2081–2092.

42 Seregard S, Lundell G, Lax I, al Trampe E, Kock E: Tumour cell proliferation after failed ruthenium plaque radiotherapy for posterior uveal melanoma. Acta Ophthalmol Scand 1997;75:148–154.

43 Verschueren KM, Creutzberg CL, Schalij-Delfos NE, Ketelaars M, Klijsen FL, Haeseker BI, Ligtenberg SM, Keunen JE, Marijnen CA: Long-term outcomes of eye-conserving treatment with ruthenium(106) brachytherapy for choroidal melanoma. Radiother Oncol 2010;95:332–338.

44 Wilson MW, Hungerford JL: Comparison of episcleral plaque and proton beam radiation therapy for the treatment of choroidal melanoma. Ophthalmology 1999;106:1579–1587.

45 Frenkel S, Hendler K, Pe'er J: Uveal melanoma in Israel in the last two decades: characterization, treatment and prognosis. Isr Med Assoc J 2009;11: 280–285.

46 Rouberol F, Roy P, Kodjikian L, Gerard JP, Jean-Louis B, Grange JD: Survival, anatomic, and functional long-term results in choroidal and ciliary body melanoma after ruthenium brachytherapy (15 years' experience with beta-rays). Am J Ophthalmol 2004;137:893–900.

47 Seregard S, aft Trampe E, Lax I, Kock E, Lundell G: Results following episcleral ruthenium plaque radiotherapy for posterior uveal melanoma. Acta Ophthalmol Scand 1997;75:11–16.

48 Jampol LM, Moy CS, Murray JG, Reynolds SM, Albert DM, Schachat AP, Diddle KR, Engstrom RE Jr, Finger PT, Hovland KR, Joffe L, Olsen KR, Wells CG, Collaborative Ocular Melanoma Study Group (COMS Group): The COMS randomized trial of iodine 125 brachytherapy for choroidal melanoma. 4. Local treatment failure and enucleation in the first 5 years after brachytherapy. Ophthalmology 2002;109:2197–2206.

49 Summanen P, Immonen I, Kivela T, Tommila P, Heikkonen J, Tarkkanen A: Visual outcome of eyes with malignant melanoma of the uvea after ruthenium plaque radiotherapy. Ophthalmic Surg Lasers 1995;26:449–460.

50 Kaiserman N, Frenkel S, Hendler K, Pe'er J: Do we need margins around the tumor base during Ru-106 brachytherapy for posterior uveal melanoma? Invest Ophthalmol Vis Sci 2010;51:ARVO E-abstract 5143.

51 Shields CL, Shields JA, Cater J, Gunduz K, Miyamoto C, Micaily B, Brady LW: Plaque radiotherapy for uveal melanoma: long-term visual outcome in 1,106 consecutive patients. Arch Ophthalmol 2000;118: 1219–1228.

52 Seregard S: Long-term survival after ruthenium plaque radiotherapy for uveal melanoma. A meta-analysis of studies including 1,066 patients. Acta Ophthalmol Scand 1999;77:414–417.

53 Kleineidam M, Guthoff R, Bentzen SM: Rates of local control, metastasis, and overall survival in patients with posterior uveal melanomas treated with ruthenium-106 plaques. Radiother Oncol 1993;28:148–156.

54 The COMS randomized trial of iodine 125 brachytherapy for choroidal melanoma. 5. Twelve-year mortality rates and prognostic factors: COMS report No 28. Arch Ophthalmol 2006;124:1684–1693.

55 Summanen P, Immonen I, Kivela T, Tommila P, Keikkonen J, Tarkkanen A: Radiation related complications after ruthenium plaque radiotherapy of uveal melanoma. Br J Ophthalmol 1996;80:732–739.
56 Gupta A, Muecke JS: Treatment of radiation maculopathy with intravitreal injection of bevacizumab (Avastin). Retina 2008;28:964–968.
57 Hogan N, Shields CL, Mashayekhi A, Salazar PF, Materin MA, O'Regan M, Shields JA: Periocular triamcinolone for prevention of macular edema after plaque radiotherapy of uveal melanoma: randomized controlled trial. Ophthalmology 2009;116:383–390.
58 Langmann A, Langmann G, Unlucerci C, Haller E: Motility disorders in brachytherapy of choroid melanomas with Ru-106 applicators. Ophthalmologe 1995;92:76–78.
59 Razzaq L, Keunen JE, Schalij-Delfos NE, Creutzberg CL, Ketalaars M, de Keizer RJ: Ruthenium plaque radiation therapy for iris and iridociliary melanomas. Acta Ophthalmol 2010, E-pub ahead of print.
60 Bartlema YM, Oosterhuis JA, Journee-deKorver JG, Tjho-Heslinga RE, Keunen JE: Combined plaque radiotherapy and transpupillary thermotherapy in choroidal melanoma: 5 years' experience. Br J Ophthalmol 2003;87:1370–1373.
61 Damato B, Foulds WS: Surgical resection of choroidal melanoma; in Schachat AP, Ryan SJ (eds): The Retina. St Louis, Mosby, 2006, pp 769–778.
62 Querques G, Bux AV, Iaculli C, Delle Noci N: Local resection versus combined local resection and plaque radiotherapy in the treatment of choroidal melanoma. Eur J Ophthalmol 2010;20:194–200.
63 Bacin F, Rozan R, Donnarieix D, Dalens H: Treatment of various malignant melanomas of the choroid using a combination of ruthenium-106 and cobalt-60. Ophthalmologie 1990;4:225–228.
64 Flühs D, Anastassiou G, Wening J, Sauerwein W, Bornfeld N: The design and the dosimetry of bi-nuclide radioactive ophthalmic applicators. Med Phys 2004;31:1481–1488.

Jacob Pe'er, MD
Department of Ophthalmology, Jonas Friedenwald Professor of Ophthalmic Research, Hadassah-Hebrew University Medical Center
PO Box 12000
Jerusalem 91120 (Israel)
Tel. +972 2 6776365, E-Mail peer@hadassah.org.il

Jager MJ, Desjardins L, Kivelä T, Damato BE (eds): Current Concepts in Uveal Melanoma.
Dev Ophthalmol. Basel, Karger, 2012, vol 49, pp 41–57

Treatment of Uveal Melanoma by Accelerated Proton Beam

Laurence Desjardins[a] · Livia Lumbroso-Le Rouic[a] · Christine Levy-Gabriel[a] · Nathalie Cassoux[a] · Remi Dendale[b] · Alexandro Mazal[c] · Sabine Delacroix[c] · Xavier Sastre[d] · Corine Plancher[e] · Bernard Asselain[e]

Departments of [a]Ophthalmology, [b]Radiotherapy, [c]Biophysics, [d]Pathology, and [e]Biostatistics, Institut Curie, Paris, France

Abstract

Proton beam irradiation of uveal melanoma has great advantages compared to brachytherapy because of the homogenous dose delivered to the tumor and the possibility of sparing normal tissue close to the tumor. We describe the technique of proton beam therapy including the surgical technique of clip positioning, the radiotherapy delivery technique and the dose administered (60 Gy cobalt relative biological effectiveness in 4 fractions). Indications of proton beam are given and the follow-up procedure is described. An inactive residual tumor scar is observed after 2–3 years. Results are given comparing the most recent series of patients treated at the Institut Curie-Orsay proton therapy center with the data published in the literature. The metastasis rate at 10 years varies between 25 and 30%. Local control is excellent. The local recurrence rate at 10 years is usually around 5%. Secondary enucleation is performed in 10–15% of patients either due to complications or local recurrence. Complications such as retinal detachment, maculopathy, papillopathy, cataract, glaucoma, vitreous hemorrhage and dryness are described. The severest complication that usually leads to secondary enucleation is neovascular glaucoma and it is encountered after irradiation of large to extra-large tumors. The toxic tumor syndrome has recently been described. It is hypothesized that the residual tumor scar may produce proinflammatory cytokines and VEGF leading to intraocular inflammation and neovascular glaucoma. Additional treatments after proton beam such as transpupillary thermotherapy, endoresection of the tumor scar or intravitreal injections of anti-VEGF may reduce the rate of these complications.

The treatment of uveal melanoma by radiotherapy was popularized by Stallard [1] who started to use cobalt plaque brachytherapy in the 1950s with encouraging results.

In the 1970s, Zimmerman et al. [2] showed that enucleation of the eye does not prevent metastatic dissemination compared to conservative management of uveal

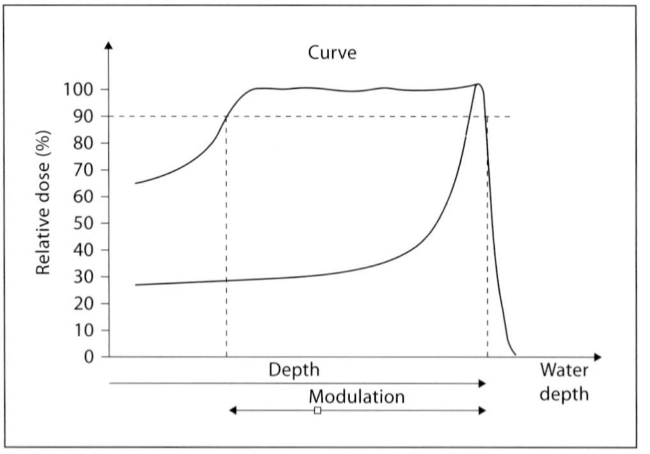

Fig. 1. Bragg peak.

melanoma. Since then treatment by brachytherapy or accelerated heavy particles has been used extensively in this disease.

Proton beam therapy of uveal melanoma has been used in Boston by Gragoudas and his group [3–5] since 1975 and in Villigen by Zografos and his group [6] since 1983. In France, two proton beam therapy centers started to treat uveal melanoma in 1991, one in Nice and one in Orsay [7, 8]. Later on, other proton beam therapy centers became available in Clatterbridge, Berlin [9] and more recently in Catana [10] in Italy. Over the past 30 years, thousands of patients have been treated in all these centers and the results have been published in the literature.

Compared to brachytherapy and to fractionated stereotactic radiotherapy [11], proton beam therapy has the advantage of delivering a homogenous dose of radiation to the entire tumor. This is due to the physical properties of accelerated proton beam which delivers the dose precisely with a sharp decrease after the delivery point called the Bragg peak [8, 12] (fig. 1). This is particularly important for uveal melanoma because this tumor is relatively radioresistant and needs a high dose of radiation [13]. That is why uveal melanoma cannot be treated by standard photon irradiation of the whole eye, as the necessary high dose of irradiation of the entire globe would induce severe damage.

Technique of Proton Beam Therapy for Uveal Melanoma

The principle of proton beam therapy derives from the physical properties of accelerated proton (the Bragg peak) [12]. The accelerated proton delivers all its energy at a precise point and there is very little radiation around and behind this point. It is then possible to deliver a high dose of radiation to the tumor while sparing the normal

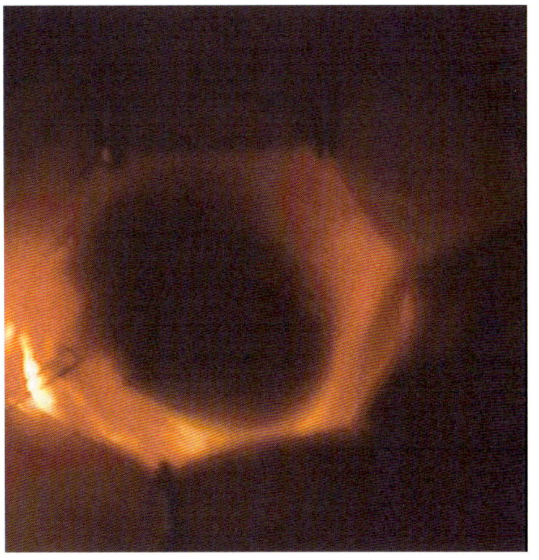

Fig. 2. Scleral transillumination.

tissue. Nevertheless it is important to keep in mind that the radiation dose at the entrance of the beam is quite high, especially for large tumors (irradiation of the lacrimal gland for tumors located at the superotemporal quadrant can induce the sicca syndrome). The sparing of normal tissue is effective on the sides and posterior part of the tumor. This sparing of normal tissue is particularly important for organs like the eye and the brain which have radiosensitive tissues [14].

The first step to treat uveal melanoma with proton beam is radiological spotting of the tumor by clip positioning. During a surgical procedure, tantalum rings of 2.5 mm in diameter with two holes in the middle are sutured to the sclera around the tumor base that has been visualized by transillumination of the globe. The conjunctiva is opened at the limbus in the area of the tumor. Traction sutures are placed on 2 or 3 rectus muscles according to the size of the tumor. Transillumination of the globe gives a precise location of the tumor base (fig. 2). If the tumor is thick, the surgeon has to move the transillumination probe (fig. 3) around the tumor to really differentiate the base of the tumor from projection of the apex. If the tumor is an amelanotic melanoma, it is usually possible to see a pigmented line at the periphery of the tumor. If the tumor is amelanotic and posterior to the equator, it is sometimes necessary to use the indirect ophthalmoscope to localize the tumor and evaluate the clip-tumor distances.

It is useful to remove the lid retractor when putting the most posterior sutures in the sclera to have enough room. In this way, sutures can be placed as far as 22 mm from the limbus. When we put the clips, we try to delineate every side of the tumor (using 4–5 clips, one on each side), but for the most posterior ones, only the anterior sides can be delineated. In this case, usually only 3 clips can be positioned close to the tumor and another clip is put more anteriorly to avoid mistakes in the planning of therapy (this is called the clip for the twist). When the clips are securely sutured to the sclera, we again use transillumination in most cases and sometimes

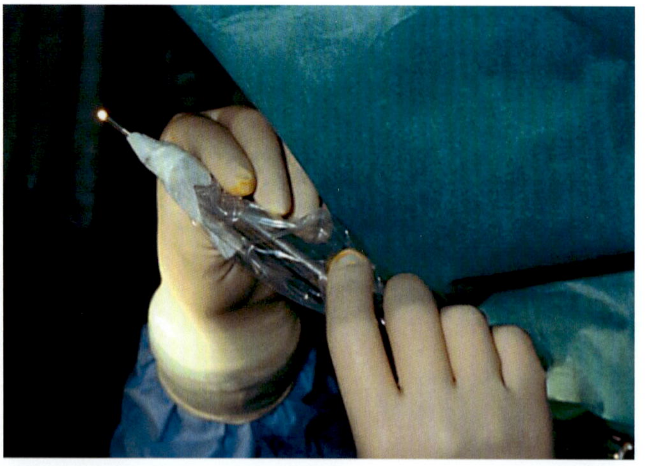

Fig. 3. Transillumination probe.

indirect ophthalmoscopy to measure the clip-tumor edge distances. We then measure the clip-limbus distances and the distances in between the clips. These measurements are critical for the subsequent computerized eye and tumor reconstruction. We then close the conjunctiva.

The patients are discharged from the hospital the following day. It is important to explain before surgery that they may have some discomfort for several days. Swelling and redness are frequent. Some patients may also complain of diplopia secondary to muscle trauma which regresses within a few days.

Proton beam therapy is planned and performed 2 weeks after surgery. The treatment technique used in Orsay has been described previously [7, 15–17]. Head immobilization during radiotherapy is ensured by a custom-made thermoplastic mask (fig. 4) associated with a bite block. All the data collected during surgery (axial length, clip measurements, tumor diameter, tumor thickness) are sent to the physicist in charge of the planning of the therapy. This planning includes the determination of the radiation field volume, which requires collaboration between ophthalmologists and the physicist. It is mandatory to have good pictures of the fundus, ultrasonography of the eye and biometry. The EYEPLAN software [18] is used for dosimetry based on three-dimensional reconstruction of the eye, including the tumor. Tumor margins are drawn by the referring center ophthalmologist and verified by the radiation oncologist. The optimal dosimetric eye position is determined according to the following criteria: total irradiation of the tumor surrounded by a safety margin of 2.5 mm and reduction of the dose delivered to the lacrimal gland, macula, optic disk and lens in decreasing order of priority. The irradiated volume is the tumor volume plus a safety margin of 2.5 mm around the tumor. This safety margin takes into account the possible imprecision of clip positioning, the risk of undetectable tumor extension or of eye movements during irradiation. If necessary, this safety margin can be increased

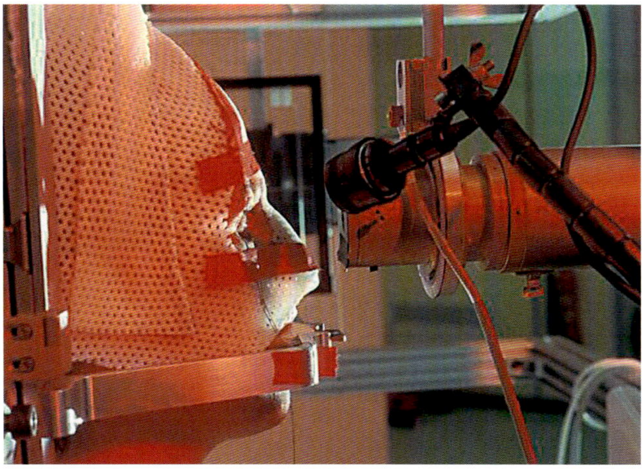

Fig. 4. Patient in treatment position.

to 3 or 3.5 mm when the localization of the tumor is difficult. This is the case when there is ciliary body involvement, nevus of Ota with diffuse pigmentation preventing the transillumination or infiltrating melanoma.

Beam characteristics are selected to place the 90% isodose 2.5 mm behind the tumor. The lateral collimator border defines the site of the 50% isodose. During treatment, patients are seated on a robot chair facing the beam and are asked to fix a small target light placed at a known angle adopted during the treatment planning process. As the beam remained horizontal the patient is moved into the treatment position. The robot chair can be moved in all directions and angles with a precision of about 0.1 mm and 0.1°. An attempt is made to keep the eyelids out of the beam's field using lid retractors. Before each irradiation session, clip positions are verified by orthogonal X-rays, and the chair is then moved to the correct position. The fixed gaze is monitored using a magnifying camera. Irradiation is suspended whenever the eye shifts from the right position and is restarted only after the eye has returned to the optimal position.

The accelerated proton beam has a low linear energy transfer and a relative biological effectiveness (RBE) of 1.1. The irradiation dose chosen by Gragoudas et al. [5] was 70 Gy RBE-weighted absorbed dose in 5 fractions. It was shown in a randomized study that there was no statistical difference in the recurrence rate between 50 and 70 Gy RBE [19]. The dose used in Lausanne, Orsay, Berlin and Catana is 60 Gy RBE in 4 fractions. In the eye, the lens is the most radiosensitive tissue. The retina and optic disk are not very radiosensitive but most of the radiation side effects are due to endothelial vascular damage causing vascular occlusions. We therefore try to spare the macula and optic disk from radiation whenever possible. The radiation antitumor effects, such as radiation tissue toxicities, depend on the total dose fractionation [20]. The biological effect of a 60-Gy dose given in 4 fractions is higher than the same 60-

Gy dose given in more fractions [20]. Despite the relatively radioresistant reputation of uveal melanoma, there are very few radioresistant cases when using this protocol.

Indications

Theoretically all uveal melanomas can be treated by accelerated proton beam. Nevertheless when large or thick tumors are treated this way, the chances of preserving vision and the globe are very low [21]. We usually do not treat tumors of more than 20 mm in diameter or tumors of more than 12 mm in thickness but there are exceptions. If the tumor is less than 15 mm in diameter, proton beam irradiation combined with endoresection of the residual tumor can give good results in terms of eye preservation, even if the tumor is thick.

Small anterior tumors can be treated by ruthenium or iodine plaques with good results for the centers that have these techniques available [22].

For large anterior tumors that are located in the superotemporal quadrant, we prefer to use iodine plaques instead of proton beam to spare the lacrimal gland.

The general status of the patient sometimes does not allow using proton beam. Comprehension and cooperation of the patient is necessary. Patients who have a psychiatric disease or mental retardation should preferentially be treated by plaque brachytherapy.

Evolution of the Tumor following Proton Beam Irradiation

The cytotoxic effect of proton beam is obtained by causing DNA damage. Cell death arises later during mitotic division [23]. Following proton beam irradiation, we observe a slow regression of the tumor in most cases and this is due to the existence of prolonged intermitotic phases in these tumors. This also explains the presence of images of mitosis in pathology reports of eyes enucleated due to complications but with no sign of tumor recurrence [24, 25]. This can also be studied using the Ki-67 index [26]. It has been shown that the number of mitotic figures possibly observed decreases with time [27]. Some of the tumor cells continue to divide several times despite lethal DNA damage from irradiation before cell death.

After proton beam irradiation, some of the tumors are changed into a flat scar (fig. 5, 6) but most of them regressed with a residual inactive tumor scar of several millimeters [28]. Like Professor Zografos [29], we have observed that amelanotic tumors tend to regress quicker and that all tumor scars are usually more pigmented than the initial tumor. In our experience, it is possible to observe an increase in thickness of some tumors of 1 or 2 mm immediately after proton beam irradiation followed by a decrease in thickness 2 or 3 months later. This transient increase in thickness may be due to edema following radiation.

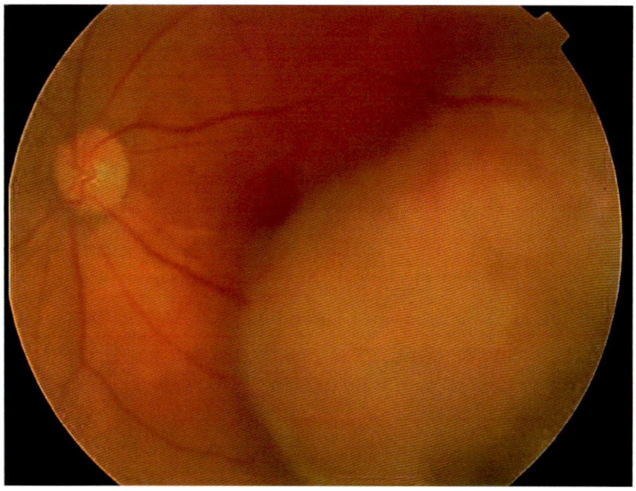

Fig. 5. Achrome melanoma before proton beam therapy.

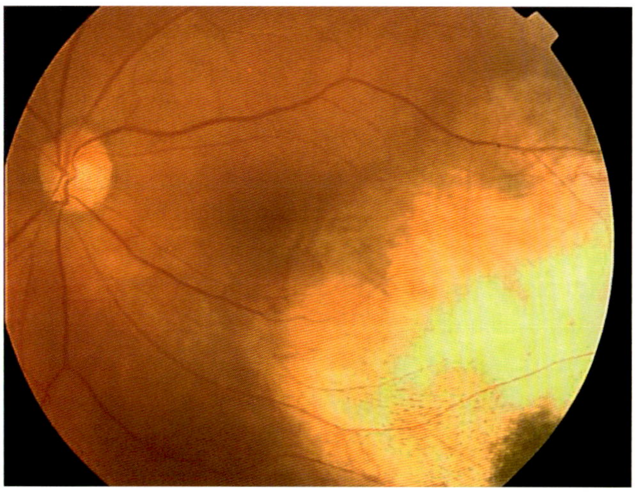

Fig. 6. Same patient after proton beam therapy.

In a few patients, it has been observed that a late rupture of Bruch's membrane with necrotic tissue can mask a recurrence.

Ultrasonography of the eye usually shows that the echogenicity of the tumor scar is higher than that of the initial tumor [30]. When Doppler imaging is performed, it shows regression and then disappearance of intratumoral vessels.

Local control of uveal melanoma is considered as the absence of tumor growth. To evaluate the tumor volume it is safe to use fundus photography for the diameter and ultrasonography for the thickness. These exams are easy to perform and can be done at each visit. Most tumors slowly decrease in size over a period of 2 years and then the residual tissue remains unchanged. It has previously been shown that a rapid decrease in tumor thickness is a sign of a more aggressive tumor with higher metastatic risk [31]. From the data collected in our database, we know that most recurrences occur during the 3 years following radiation therapy. There are 3 types of recurrences [32]: (1) recurrence at the border is probably due to an insufficient radiation dose at the border following errors in the planning of the therapy or in the irradiation delivery; (2) recurrence in the inferior periphery, distant from the initial tumor, may be due to migration of tumor cells in retinal detachment; careful examination of the inferior periphery should be made at each postirradiation control; (3) recurrence in the field of radiation due to radioresistance may be responsible for the late recurrences (more than 10 years after irradiation) and is rarely observed in this disease.

Follow-Up of Patients

Patients at the Curie Institute are checked every 6 months for 2 years and then once a year for 10 years. They are advised to see their local ophthalmologist in between. Ultrasonography of the liver is performed routinely every 6 months. Because of an increasing number of patients in the follow-up clinics, we have decided more recently to see the patients less often as soon as the time where recurrences occur has passed: after 3 years of follow-up, we see them only twice, i.e. at 5 and 10 years. However, we still advise them to see their local ophthalmologist every 6 months.

Most ophthalmologists dealing with uveal melanoma have a database. At the Curie Institute, the database was started in 1983 [33] when uveal melanomas were treated by cobalt plaque brachytherapy. Information concerning the initial findings (tumor dimension and location, ciliary body involvement, retinal detachment, iris color), local treatment, and the follow-up data (survival, metastasis and secondary enucleation, thickness of the tumor scar and ocular side effects) is collected prospectively. In 2008, with financial support of the national cancer institute (INCA), the database was extended to all centers treating uveal melanoma in France. To date, 5,724 patients have been included in this French database.

Results of Proton Beam Therapy

We recently analyzed patients treated for uveal melanoma at the Curie Institute and at the Orsay proton therapy center between October 1991 and October 2006 (minimum follow-up 4 years). I shall give you the results of this study (not yet published data) and compare these results to the results published in the literature. Survival rates were determined using Kaplan-Meier estimates. Prognostic parameters were analyzed using the logrank test (univariate analysis) and Cox model (multivariate analysis).

Initial Findings

The study included 2,413 treated patients with a median age at diagnosis of 61 years ranging from 14 to 95 years. There were equal numbers in both genders (sex ratio 0.99). For 1,207 patients, we obtained information on the iris color, which was blue in 52%, green in 14% and brown in 31% with green and brown in 3%.

The tumor was localized anterior to the equator in 3%, crossing the equator in 39% and posterior to the equator in 57%. The ciliary body was involved in 8% of cases. Median tumor diameter was 13.7 mm and median tumor thickness was 4.7 mm. According to the new TNM classification [34], there were 414 T1, 398 T2, 1,172 T3 and 420 T4 tumors.

Retinal Detachment

A retinal detachment was present at diagnosis in 26% of cases. The tumor was touching the optic disk in 7.4% of cases. There was extrascleral extension in 1.3% of cases. (We showed in a previous study that eyes presenting with localized extrascleral extension can be treated conservatively with proton beam with good results [35].) Initial visual acuity was less than 20/200 in 14%, from 20/200 to 20/40 in 49% and better than 20/40 in 37%. The median follow-up of the patients was 98 months (range 1–175 months).

Survival

When we looked at overall survival, 30% of the patients were dead. The 5-year survival rate was 79%. In a multivariate analysis, the binary risk factors that significantly correlated with survival were male versus female gender [hazard ratio (HR) = 1.35, $p < 0.001$), age at diagnosis ≥60 years versus <60 years (HR = 2.67, $p < 0.0001$), tumor diameter ≥15 mm versus <15 mm (HR = 2.02, $p < 0.0001$), tumor thickness >5 mm versus ≤5 mm (HR = 1.52, $p < 0.0001$), retinal detachment (HR = 1.27, $p = 0.006$) and tumor location anterior or on equator versus posterior (HR = 1.27, $p = 0.008$).

Metastases

We observed that 22% of patients developed metastases. The 5-year metastasis rate was 18.5% and the 10-year rate was 26.6%. Significant risk factors for the development of metastases in a multivariate analysis are: retinal detachment (HR = 1.48, $p < 0.001$), tumor location anterior or on equator versus posterior (HR = 1.73, $p < 0.0001$), tumor diameter ≥15 mm versus <15 mm (HR = 2, $p < 0.0001$), tumor thickness >5 mm

Table 1. Metastatic rate after proton beam therapy for uveal melanoma

Reference; year	Patients, n	5-year metastatic rate, %	10-year metastatic rate, %
Egan et al. [36]; 1998	1,541	15	25
Courdi et al. [37]; 1999	538	23 (7.5 years)	
Egger et al. [39]; 2001	2,258	16	31
Dendale et al. [15]; 2006	1,406	19	
Caujolle et al.[38]; , 2010	886	13	24
CPO Orsay [unpublished data]; 2010	2,413	19	27

versus ≤5 mm (HR = 1.62, $p < 0.0001$), age at diagnosis ≥60 years versus <60 years (HR = 1.49, $p < 0.0001$) and local recurrence ($p < 0.0001$). Table 1 gives the metastatic rate at 5 and 10 years after treatment of uveal melanoma by accelerated proton beam as published in the literature [29, 36–38]. In all these series, the 10-year metastatic rate is between 25 and 30%. This is less than the 10-year mortality rate observed after enucleation but an explanation for this is that the tumors treated by proton beam are smaller, so their survival prognosis is better.

Local Control

With regard to local control, 4.3% of patients developed a local recurrence. Ten percent of these recurrences were distant recurrences due to migration of tumor cells in retinal detachment. The 5-year local recurrence rate was 4% and the 10-year rate was 5%. Multivariate analysis shows that a tumor diameter ≥15 mm versus <15 mm (HR = 2.97, $p < 0.0001$) and retinal detachment (HR = 1.65, $p = 0.02$) are significantly correlated with local recurrence. Table 2 gives the percentage of recurrence in published series after treatment of uveal melanoma by proton beam [15, 32, 38–40]. Local control is most of the time excellent, being around 95% at 10 years. Local recurrence occurs mostly during the first 3 years of treatment but can also occur more than 10 years after treatment. Thus follow-up should be frequent during the first 3 years with fundus exam every 3 months, and an annual fundus exam should be maintained for life.

In case of local recurrence, treatment depends on the status of the globe and the size and position of the recurrence. In our center, 75% of recurrences are treated by enucleation. A second treatment by proton beam can be performed in selected cases but the patient has to be aware that a second irradiation increases the risk of radiation complications and atrophy of the globe. Treatment of the recurrence by transpupillary thermotherapy (TTT) can sometimes be performed in case of very limited recurrence.

Table 2. Local recurrence rate after proton beam therapy for uveal melanoma

Reference; year	Cases, n	5-year rate, %	10-year rate, %
Gragoudas et al. [32]; 1992	1,077	3	
Egger et al. [39]; 2001	2,435	4 (1989–1993), 1.1 (1993–2001)	
Gragoudas et al. [40]; 2002	1,992	3	4
Dendale et al. [15]; 2006	1,406	4	
Caujolle et al. [38]; 2010	886	6	8
CPO Orsay [unpublished data]; 2010	2,413	4 (98 months)	

Table 3. Secondary enucleation after proton beam therapy for uveal melanoma

Reference; year	Patients, n	5-year rate, %	10-year rate, %
Egan et al. [36]; 1998	1,541	8	11
Egger et al. [41]; 2003	2,645	11	14
CPO Orsay [unpublished data]; 2010	2,413	10 (98 months)	

Secondary Enucleation

A total of 256 eyes (10%) underwent secondary enucleation. With a median follow-up of 98 months, 181 patients were enucleated due to complications and 75 due to recurrence. Multivariate analysis of eyes enucleated due to complications showed that the significant risk factors were tumor thickness >5 mm versus ≤5 mm (HR = 5, $p < 0.0001$) and, to a lesser extent, tumor diameter ≥15 mm versus <15 mm (HR = 1.64, $p = 0.003$). It is interesting to note that in this cohort of patients the tumor diameter and the tumor thickness are of equal importance for survival and metastases, but for local recurrence the diameter seems more important and for secondary enucleation the thickness is more significant. The reason for enucleation in patients enucleated due to complications was glaucoma in most cases or a painful nonseeing eye. Table 3 gives the secondary enucleation rates after proton beam irradiation as published in the literature [36, 41].

Complications

With regard to complications, we saw that with 98 months of follow-up, patients had loss of lashes in 12%, retinal detachment in 8.5%, and glaucoma in 23.4%. Forty-eight percent of the patients had lens opacities or cataract and 15% had had cataract surgery. Dry eye was observed in 6%, keratitis in 7%, vitreous hemorrhage in 8%, optic neuropathy in 18%, active inflammation in 1% and maculopathy in 37%.

Retinal Detachment

Retinal detachment, if present at diagnosis, usually had a tendency to increase during the first months after proton beam irradiation, especially with large tumors. It can also appear after irradiation even if it was not present at diagnosis. It usually disappears within 3 years. It has recently been shown by Professor Zografos, using wide-angle angiography, that if there are more than 3 quadrants of retinal detachment, there is a high risk of peripheral ischemia and neovascular glaucoma (EVER meeting 2010).

Neovascular Glaucoma

Neovascular glaucoma, glaucoma and inflammation are possibly linked to what is called 'the toxic tumor syndrome'. It is thought that the irradiated tumor scar, especially for thick tumors, is responsible for the secretion of cytokines leading to inflammation [42] and of VEGF leading to neovascularization of the anterior segment [43–45]. The ischemic retina may also be the origin of developing neovascularization. Neovascular glaucoma is a major reason for enucleation after proton beam irradiation of uveal melanoma. It usually appears between 2 and 5 years after radiation therapy. The size of the tumor and the persistence of retinal detachment are the most frequent predisposing factors [46]. Associated therapies can be applied after proton beam to try to prevent neovascular glaucoma. When neovascular glaucoma is present, it is important to control the pressure to avoid pain, enucleation and evolution to globe atrophy. The use of antiglaucomatous and anti-inflammatory medication can be useful. Filtering surgery is most often very disappointing and excessive cryotherapy to the ciliary body usually leads to severe atrophy of the globe. More recently [47], intravitreal injections of anti-VEGF have been used and are often very useful allowing regression of the neovascularization and improvement of the status of the globe. Nevertheless, if the neovascular glaucoma is associated with cataract, synechia and pain and the fundus cannot be observed any more due to opaque media, it may sometimes be necessary to enucleate the eye.

Cataract

Cataract is the most frequent complication in most series [6, 48]. The radiosensitivity of the lens is high [49]. As irradiation by proton beam is localized to one quadrant, sector cataract is frequently observed in the irradiated quadrant. Cataracts due to radiation are more likely to be posterior capsular opacities while nuclear cataracts are not standardly linked to radiation but can be linked to the age of the patient. If the tumor is large and if the irradiation dose to the lens is important, most patients develop total white cataracts. In this case, we always recommend cataract surgery to be able to examine the fundus, even if no visual benefit can be obtained because of a macular or peripapillary location of the melanoma. It is usually mandatory to wait 2 years after proton beam irradiation prior to performing cataract surgery in order to observe a good regression of the tumor. Surgeons should be aware that the capsule is sometimes difficult to clean and that inflammatory reactions after surgery can be severer than for a standard cataract surgery.

Optic Neuropathy

Optic neuropathy is not due to direct toxicity of the radiation to the nerve fibers but to vascular endothelial cell damage causing vascular occlusions [50, 51]. It usually appears between 18 months and 2 years after the irradiation with a high risk (around 50%) if the optic disk receives 100% of the dose. Patients may complain of sudden loss of vision and the fundus exam shows optic disk edema with hemorrhages and cotton wool spots. It is not known if anticoagulant medication could have a preventive effect. The treatment is difficult and most patients lose vision because of optic atrophy. If the angiograms show associated retinal ischemia, we recommend performing a panretinal photocoagulation to avoid neovascular glaucoma. I have observed a case of a monophthalmic patient who was cured by intravenous heparin and subsequent oral anticoagulant medication for more than 10 years keeping 20/20 vision.

Radiation Maculopathy

Radiation maculopathy is also due to toxicity of the radiation to the vascular supply of the macula. It includes microaneurysms, cotton wool spots, telangiectasia, retinal hemorrhages, exudates and cystoid macular edema. Angiography shows areas of nonperfusion, microaneurysms and exudation. Radiation maculopathy can occur within 3 years following radiation, sometimes later. The visual loss depends on the severity of the maculopathy. Because of the small size of macular capillaries, radiation maculopathy can occur with small doses of radiation. Slight maculopathies can be seen several years after proton beam therapy of hemangiomas with a dose of 20 Gy when the macula is in the radiation field. The treatment of radiation maculopathy is disappointing. Intravitreal injections of corticosteroids or anti-VEGF drugs may improve the visual acuity in some patients but only for a short period of time. It is important to note that some patients treated by proton beam for uveal melanoma, especially large tumors, have macular alterations and cystoid macular edema even if the macula has not received radiation. This is probably due to other mechanisms like proinflammatory cytokines production and intraocular inflammation.

Vitreous Hemorrhage

Vitreous hemorrhage is less frequent with proton beam than with plaque brachytherapy according to a prospective randomized study comparing these two methods of treatment [52]. Vitreous hemorrhage usually spontaneously resolves in a few weeks. If this is not the case, a vitrectomy procedure can be proposed in selected cases. Recurrence of the vitreous hemorrhage is possible but not standard. Recurrent vitreous hemorrhage can be an indication to perform enucleation as it is not possible to see the fundus and ultrasonography is difficult to interpret because of the echoes of the hemorrhage.

Visual Acuity

In our series of 2,413 patients with a median follow-up of 98 months, 58% of patients had a last noticed visual acuity of less than 20/200, while 42% had at least 20/200 and 21% more than 20/40. Gragoudas et al. [19] showed in a randomized study that visual acuity was not influenced by the total delivered dose. After using 50 or 70 Gy, the percentage of patients with a visual acuity of 20/200 or more was 56% for 50 Gy and 54% for 70 Gy. The main factors responsible for visual loss after proton beam therapy are the tumor volume and the tumor location. Patients with tumors located close to the macula or optic disk have a high risk of having radiation maculopathy or papillopathy with loss of vision. Patients with large tumors are at a high risk of developing inflammation and neovascular glaucoma.

Additional Treatments

Additional treatments are sometimes used after proton beam irradiation in order to improve globe preservation and if possible improve final visual results. For this purpose, we have used additional TTT and more recently surgical endoresection of the tumor scar and intravitreal injections of anti-VEGF drugs. We have used additional TTT after proton beam irradiation in two different circumstances. First we have used it to reduce the radiation dose to the macula if the tumor is close to it. Studies demonstrate [32, 39] that reduction of the security margin induces a higher risk of tumor recurrence. We included 10 patients in a pilot study where we reduced the security margin in the macular area from 2.5 to 1.5 mm, using TTT to the posterior margin of the tumor to avoid recurrences. All patients had a final good visual acuity except 1 who developed a branch vein occlusion. None of the patients presented with a recurrence. Since then, we have been using this technique in selected cases.

The second protocol was a randomized study using 3 applications of TTT after proton beam irradiation to try to avoid neovascular glaucoma [53]. We believed that applying TTT to the surface of the tumor scar might prevent the secretion of cytokines and VEGF. One hundred and fifty-one patients were randomized and we showed that the group of patients receiving additional TTT had less secondary enucleation ($p < 0.02$). Applying TTT to the surface of the tumor scar is one way of preventing the tumor scar to be toxic. More recently, Bechrakis et al. [54, 55] described the association of surgical endoresection of the tumor scar after proton beam radiation. Endoresection was first described by Damato et al.

[56]. Some published cases showed that a massive intraocular recurrence can be observed if endoresection is preformed without any associated irradiation therapy [57], and it seems to us safer to use it after proton beam irradiation, though some authors use it as a single procedure [58]. Indications to perform endoresection after proton beam irradiation are tumors that are less than 15 mm in diameter and more than 8 mm in thickness, not involving the optic disk or ciliary body. We have used this technique in 50 patients over the last 2 years with promising results.

In case of neovascular glaucoma, intravitreal or intracameral injections of bevacizumab (Avastin®) can help as it usually induces regression of the neovascularization and decrease of ocular pressure [47].

Finally, it is important to have different options of additional treatments in case of large tumors. Some patients may benefit from TTT if they cannot have surgery.

Conclusion

Proton beam therapy of uveal melanoma is one of the best treatment modalities allowing excellent tumor control and good survival. For patients with large tumors, additional work is necessary to continue to improve globe preservation, visual acuity and quality of life. Combined treatment modalities (radiotherapy + surgery, radiotherapy + laser, radiotherapy + intravitreal injections of anti-VEGF) are promising but need to be evaluated in a prospective way.

Acknowledgement

We thank all the physicists who have contributed to eye tumor treatments at the Institut Curie-Orsay proton therapy center: C. Nauraye, C. Mabit, I. Pasquié, and R. Ferrand.

References

1 Stallard HB: Radiotherapy for malignant melanoma of the choroid. Br J Ophthalmol 1966;50:147–155.
2 Zimmerman L, McLean I, Foster W: Does enucleation of the eye containing a malignant melanoma prevent or accelerate the dissemination of tumour cells ? Br J Ophthalmol 1978;62:420–425.
3 Gragoudas ES, Seddon JM, Egan K, et al: Long-term results of proton beam irradiated uveal melanomas. Ophthalmology 1987;94:349–353.
4 Gragoudas E, Seddon J, Egan K, et al: Proton Irradiation of Uveal Melanomas: The First 1,000 Patients. Amsterdam, Kugler Publications, 1991.
5 Gragoudas ES, Goitein M, Verhey L, Munzenreider J, Suit HD, Koehler A: Proton beam irradiation. An alternative to enucleation for intraocular melanomas. Ophthalmology 1980;87:571–581.
6 Zografos L, Bercher L, Egger E, et al: Treatment of eye tumors by accelerated proton beams. 7 years experience. Klin Monatsbl Augenheilkd 1992;200: 431–435.
7 Desjardins L, Levy C, d'Hermies F, et al: Initial results of proton therapy in choroidal melanoma at the d'Orsey Center for Proton Therapy; the first 464 cases. Cancer Radiother 1997;1:222–226.

8 Nauraye C, Mazal A, Delacroix S, Bridier A, Chavaudra J, Rosenwald JC: An experimental approach to the design of a scattering system for a proton therapy beam line dedicated to ophthalmological applications. Int J Radiat Oncol Biol Phys 1995;32:1177–1183.
9 Hocht S, Bechrakis NE, Nausner M, et al: Proton therapy of uveal melanomas in Berlin. 5 years of experience at the Hahn-Meitner Institute. Strahlenther Onkol 2004;180:419–424.
10 Spatola C, Privitera G, Raffaele L, et al: Clinical application of proton beams in the treatment of uveal melanoma: the first therapies carried out in Italy and preliminary results (CATANA Project). Tumori 2003;89:502–509.
11 Weber DC, Bogner J, Verwey J, et al: Proton beam radiotherapy versus fractionated stereotactic radiotherapy for uveal melanomas: a comparative study. Int J Radiat Oncol Biol Phys 2005;63:373–384.
12 Gragoudas ES: The Bragg peak of proton beams for treatment of uveal melanoma. Int Ophthalmol Clin 1980;20:123–133.
13 Habermalz HJ, Fischer JJ: Radiation therapy of malignant melanoma: experience with high individual treatment doses. Cancer 1976;38:2258–2262.
14 Mazal A, Delacroix S, Arianer J, et al: Protontherapy: physical and technical basis. Bull Cancer Radiother 1996;83:230–246.
15 Dendale R, Lumbroso-Le Rouic L, Noel G, et al: Proton beam radiotherapy for uveal melanoma: results of Curie Institut-Orsay proton therapy center (ICPO). Int J Radiat Oncol Biol Phys 2006;65: 780–787.
16 Hamrouni Z, Levy C, Lumbroso L, et al: Results of treating uveal melanoma with proton beam radiation: 10-year follow-up. J Fr Ophtalmol 2005;28:833–839.
17 Lumbroso L, Levy C, Plancher C, et al: Results of proton beam irradiation for treatment of choroidal melanoma. J Fr Ophtalmol 2002;25:290–297.
18 Goitein M, Miller T: Planning proton therapy of the eye. Med Phys 1983;10:275–283.
19 Gragoudas ES, Lane AM, Regan S, et al: A randomized controlled trial of varying radiation doses in the treatment of choroidal melanoma. Arch Ophthalmol 2000;118:773–778.
20 Wollensak G, Zografos L, Perret C, Egger E, Fritz-Niggli H: Experimental study on the fractionation schedule for proton irradiation of uveal melanoma. Graefes Arch Clin Exp Ophthalmol 1990;228:562–568.
21 Conway RM, Poothullil AM, Daftari IK, Weinberg V, Chung JE, O'Brien JM: Estimates of ocular and visual retention following treatment of extra-large uveal melanomas by proton beam radiotherapy. Arch Ophthalmol 2006;124:838–843.
22 Desjardins L, Lumbroso L, Levy C, et al: Treatment of uveal melanoma with iodine 125 plaques or proton beam therapy: indications and comparison of local recurrence rates. J Fr Ophtalmol 2003;26:269–276.
23 Lichter AS, Lawrence TS: Recent advances in radiation oncology. N Engl J Med 1995;332:371–379.
24 Crawford JB, Char DH: Histopathology of uveal melanomas treated with charged particle radiation. Ophthalmology 1987;94:639–643.
25 Saornil MA, Egan KM, Gragoudas ES, Seddon JM, Walsh SM, Albert DM: Histopathology of proton beam-irradiated vs enucleated uveal melanomas. Arch Ophthalmol 1992;110:1112–1118.
26 Chiquet C, Grange JD, Ayzac L, Chauvel P, Patricot LM, Devouassoux-Shisheboran M: Effects of proton beam irradiation on uveal melanomas: a comparative study of Ki-67 expression in irradiated versus non-irradiated melanomas. Br J Ophthalmol 2000; 84:98–102.
27 Gragoudas ES, Egan KM, Saornil MA, Walsh SM, Albert DM, Seddon JM: The time course of irradiation changes in proton beam-treated uveal melanomas. Ophthalmology 1993;100:1555–1560.
28 Wilkes SR, Gragoudas ES: Regression patterns of uveal melanomas after proton beam irradiation. Ophthalmology 1982;89:840–844.
29 Zografos L: Tumeurs intraoculaires. Paris, Masson, 2002.
30 Boudinet M, Berges O, Le Huerou JY, Lumbroso-Le Rouic L, Desjardins L, Laugier P: Quantitative echography in the follow-up of patients treated with proton-beam irradiation for primary choroidal melanomas. Ultrasound Med Biol 2007;33:1046–1056.
31 Kaiserman I, Anteby I, Chowers I, Blumenthal EZ, Kliers I, Pe'er J: Post-brachytherapy initial tumour regression rate correlates with metastatic spread in posterior uveal melanoma. Br J Ophthalmol 2004;88:892–895.
32 Gragoudas ES, Egan KM, Seddon JM, Walsh SM, Munzenrider JE: Intraocular recurrence of uveal melanoma after proton beam irradiation. Ophthalmology 1992;99:760–766.
33 Desjardins L, Levy-Gabriel C, Lumbroso-Lerouic L, et al: Prognostic factors for malignant uveal melanoma. Retrospective study on 2,241 patients and recent contribution of monosomy-3 research. J Fr Ophtalmol 2006;29:741–749.
34 Finger PT: The 7th edition AJCC staging system for eye cancer: an international language for ophthalmic oncology. Arch Pathol Lab Med 2009;133:1197–1198.
35 Bellmann C, Lumbroso-Le Rouic L, Levy C, et al: Uveal melanoma: management and outcome of patients with extraocular spread. Br J Ophthalmol 2010;94:569–574.

36 Egan KM, Ryan LM, Gragoudas ES: Survival implications of enucleation after definitive radiotherapy for choroidal melanoma: an example of regression on time-dependent covariates. Arch Ophthalmol 1998;116:366–370.

37 Courdi A, Caujolle JP, Grange JD, et al: Results of proton therapy of uveal melanomas treated in Nice. Int J Radiat Oncol Biol Phys 1999;45:5–11.

38 Caujolle JP, Mammar H, Chamorey E, Pinon F, Herault J, Gastaud P: Proton beam radiotherapy for uveal melanomas at nice teaching hospital: 16 years' experience. Int J Radiat Oncol Biol Phys 2010;78:98–103.

39 Egger E, Schalenbourg A, Zografos L, et al: Maximizing local tumor control and survival after proton beam radiotherapy of uveal melanoma. Int J Radiat Oncol Biol Phys 2001;51:138–147.

40 Gragoudas E, Li W, Goitein M, Lane AM, Munzenrider JE, Egan KM: Evidence-based estimates of outcome in patients irradiated for intraocular melanoma. Arch Ophthalmol 2002;120: 1665–1671.

41 Egger E, Zografos L, Schalenbourg A, et al: Eye retention after proton beam radiotherapy for uveal melanoma. Int J Radiat Oncol Biol Phys 2003;55: 867–880.

42 Ly LV, Bronkhorst IH, van Beelen E, et al: Inflammatory cytokines in eyes with uveal melanoma and relation with macrophage infiltration. Invest;51:5445–5451.

43 Boyd SR, Gittos A, Richter M, Hungerford JL, Errington RD, Cree IA: Proton beam therapy and iris neovascularisation in uveal melanoma. Eye (Lond) 2006;20:832--836.

44 Boyd SR, Tan D, Bunce C, et al: Vascular endothelial growth factor is elevated in ocular fluids of eyes harbouring uveal melanoma: identification of a potential therapeutic window. Br J Ophthalmol 2002;86: 448–452.

45 Missotten GS, Notting IC, Schlingemann RO, et al: Vascular endothelial growth factor a in eyes with uveal melanoma. Arch Ophthalmol 2006;124:1428–1434.

46 Foss AJ, Whelehan I, Hungerford JL, et al: Predictive factors for the development of rubeosis following proton beam radiotherapy for uveal melanoma. Br J Ophthalmol 1997;81:748–754.

47 Vasquez LM, Somani S, Altomare F, Simpson ER: Intracameral bevacizumab in the treatment of neovascular glaucoma and exudative retinal detachment after brachytherapy in choroidal melanoma. Can J Ophthalmol 2009;44:106–107.

48 Gragoudas ES, Egan KM, Walsh SM, Regan S, Munzenrider JE, Taratuta V: Lens changes after proton beam irradiation for uveal melanoma. Am J Ophthalmol 1995;119:157–164.

49 Merriam GR Jr, Focht EF: A clinical study of radiation cataracts and the relationship to dose. Am J Roentgenol Radium Ther Nucl Med 1957;77:759–785.

50 Hopewell JW: Radiation injury to the central nervous system. Med Pediatr Oncol 1998;Suppl 1:1–9.

51 Coderre JA, Morris GM, Micca PL, et al: Late effects of radiation on the central nervous system: role of vascular endothelial damage and glial stem cell survival. Radiat Res 2006;166:495–503.

52 Char DH, Quivey JM, Castro JR, Kroll S, Phillips T: Helium ions versus iodine 125 brachytherapy in the management of uveal melanoma. A prospective, randomized, dynamically balanced trial. Ophthalmology 1993;100:1547–1554.

53 Desjardins L, Lumbroso-Le Rouic L, Levy-Gabriel C, et al: Combined proton beam radiotherapy and transpupillary thermotherapy for large uveal melanomas: a randomized study of 151 patients. Ophthalmic Res 2006;38:255–260.

54 Bechrakis NE, Hocht S, Martus P, Kreusel KM, Heese J, Foerster MH: Endoresection following proton beam irradiation of large uveal melanomas. Ophthalmologe 2004;101:370–376.

55 Bechrakis NE, Foerster MH: Neoadjuvant proton beam radiotherapy combined with subsequent endoresection of choroidal melanomas. Int Ophthalmol Clin 2006;46:95–107.

56 Damato B, Groenewald C, McGalliard J, Wong D: Endoresection of choroidal melanoma. Br J Ophthalmol 1998;82:213–318.

57 Hadden PW, Hiscott PS, Damato BE: Histopathology of eyes enucleated after endoresection of choroidal melanoma. Ophthalmology 2004;111:154–160.

58 Karkhaneh R, Chams H, Amoli FA, et al: Long-term surgical outcome of posterior choroidal melanoma treated by endoresection. Retina 2007;27:908–914.

Laurence Desjardins
Department of Ophthalmology, Institut Curie
26, rue d'Ulm
FR–75005 Paris (France)
E-Mail laurence.desjardins@curie.net

Jager MJ, Desjardins L, Kivelä T, Damato BE (eds): Current Concepts in Uveal Melanoma.
Dev Ophthalmol. Basel, Karger, 2012, vol 49, pp 58–65

Stereotactic Photon Beam Irradiation of Uveal Melanoma

Martin Zehetmayer

Department of Ophthalmology and Optometry, Medical University of Vienna, Vienna, Austria

Abstract

Stereotactic photon beam irradiation has been under clinical investigation for the treatment of uveal melanoma for over 15 years. Single-fraction stereotactic radiosurgery (SRS) is usually done with a gamma knife as well as more recently with a cyberknife. The therapeutic single dose has been reduced to as low as 35 Gy over the past few years without reduction in tumor control. Doses of 40 Gy delivered at the 50% isodose result in good local tumor control and acceptable toxicity. Since radiobiological studies indicate a possible advantage of hypofractionated treatment over a single very large fraction to sterilize uveal melanoma cell lines, fractionated stereotactic radiotherapy (SRT) has gained additional interest. Besides increased tumor control, toxicity should theoretically be reduced by fractionation. Linear accelerators (LINAC) have the advantage of a feasible fractionation. Most LINAC studies employ a hypofractionated scheme of 4–5 fractions and total doses between 50 and 70 Gy. The efficacy of SRT for uveal melanoma has been proven in different studies with local tumor control rates reported over 90%, 5 and 10 years after treatment. Radiogenic side effects after SRT are reported similarly to other forms of radiotherapy, with cataract development, radiation retinopathy, opticopathy and neovascular glaucoma being responsible for the majority of secondary vision losses and secondary enucleations. Overall, stereotactic photon beam radiotherapies (SRS and SRT) are considered effective treatment modalities for uveal melanoma, with promising late tumor control and toxicity rates. Additional studies and longer follow-up are indicated to finally confirm optimal treatment modalities.

Uveal melanoma is the most common primary malignant intraocular tumor in adults. Although this tumor was traditionally believed to be radioresistant, nowadays radiotherapy is the most convenient form of conservative treatment for this disease. Besides brachytherapy, teletherapy with proton and helium ion charged-particle irradiation has successfully been used for more than three decades.

Stereotactic irradiation techniques using photon beams have been under investigation for well over a decade for the treatment of uveal melanoma, with the first clinical reports of stereotactic external beam irradiation having been published over

15 years ago [1–4]. The word 'stereotactic' is of Greek origin (*στερεός* 'solid, firm, spatial', *τάσσω* 'alignment, arrangement, ordering'). Regarding terminology, 'stereotactic radiosurgery (SRS)' refers to a treatment consisting of a single very large fraction, whereas 'fractionated stereotactic radiotherapy (SRT)' refers to a treatment using multiple fractions. A range of studies concerning stereotactic irradiation of uveal melanoma have been published, ranging from radiosurgical strategies with different doses (80–35 Gy) to hypofractionated radiotherapeutic modalities (usually 4–5 fractions and total doses between 50 and 70 Gy). A major advantage of stereotactic photon beam radiosurgery/radiotherapy is the fact that usually no preliminary surgery is needed to localize the tumor, but tumor border delineation will be done by CT or MRI or, most recently, by image fusion, a combination of both. The tumor volume is depicted within a three-dimensional array facilitating treatment beam alignment.

Stereotactic photon beam irradiation can be performed using a gamma knife, a linear accelerator (LINAC) or a cyberknife.

Both SRS and SRT require techniques for ocular immobilization. These range from retrobulbar anesthesia, bridle sutures through the rectus muscles to retain the eye in position, to more sophisticated solutions, like vacuum suction cups for the gamma knife and TV camera systems with computer-assisted eye tracking and automatic gating for the LINAC.

In the following sections give an overview and discuss the different stereo- tactic photon beam treatment forms for uveal melanoma, as well as clinical results reported in the literature.

Gamma Knife

Originally introduced to treat intracranial tumors, the gamma knife was invented in the 1960s by the Swedish neurosurgeon Lars Leksell. The gamma knife itself consists of 201 fixed cobalt-60 sources arranged in 5 concentric rings. These cobalt sources have a cylindrical form and have a specific activity of 1 TBq. To achieve irradiation of the selected target volume, the radiation emitted by the cobalt sources is aimed to a common focal point by a primary collimator. Secondary collimation using a collimating helmet allows the adjustment of the gamma ray beam focus size. Thereby the radiation beams of the 201 cobalt sources are targeted towards the focus. Each individual beam has relatively low energy, so the radiation has little effect on intervening tissue and is concentrated on the tumor itself. Thus irradiation of the selected volume is achieved while the surrounding tissue is exposed to irradiation to a much lesser degree. Experimental results of the successful treatment of melanoma in a rabbit eye model were reported by Rand et al. [5] in 1987. Long-term results of SRS for uveal melanoma using a gamma knife show satisfying tumor control with rates around 90% [6–11]. However, secondary side effects are common. Haas et al. [12] found radiation retinopathy in as many as 84% of their patients who underwent SRS with a median marginal dose of 50 Gy at a Leksell gamma knife unit. Severe forms of radiation opticopathy were found in 8.6–20% of anterior and posterior melanomas [6, 9–11]. Also, high rates of neovascular glaucoma occurrence were reported by many centers – i.e., rates of up to 48%, depending on the total dose applied, tumor size and location [6, 7]. Recent studies report lower incidences of radiogenic side effects by lowering the total irradiation dose to around 40 Gy and even 35 Gy at the 50% isodose [7, 11].

One main issue with gamma knife treatments in general is fixation – due to a rigid, invasive stereotactic frame needed for target fixation, the possibility of fractionation is thereby limited. Although there are successful studies showing the benefits of a fractionated irradiation scheme in the treatment of uveal melanoma using the gamma knife [13], radiosurgery has remained the most widespread treatment form for this disease with this device. In the study mentioned above, our Vienna group prospectively evaluated local tumor control and morbidity after 1–3 fractions of stereotactic external beam irradiation in 62 selected patients with uveal melanoma, unsuitable for ruthenium-106 brachytherapy or local resection. The mean initial tumor height was 7.8 ± 2.8 mm. With the Leskell gamma knife stereotactic external beam irradiation, 41 patients (66%) were irradiated with 2 equal fractions of 35, 30 or 25 Gy/fraction, 14 patients (22%) were treated with 3 fractions of 15 Gy each, and 7 patients (11%) with small tumor volumes below 400 mm^3 were treated with 1 fraction of 45 Gy. The mean total dose was 54 ± 8 Gy. The median follow-up was 28 months. Local tumor control was achieved in 98% and tumor height reduction in 97%. Seven patients developed metastases (11%). Secondary enucleation was performed in 8 eyes (13%). Morbidity was significant in tumors exceeding 8 mm in initial height and in smaller tumors, morbidity was acceptable and comparable to literature data. In a stepwise multiple Cox model, the high-dose volume irradiated with more than 10 Gy/fraction was the strongest risk factor for radiation-induced inflammatory reactions. We concluded that especially fractionated stereotactic irradiation has potential as an eye-preserving treatment for uveal melanoma, and we initiated studies with LINAC fractionated SRT.

Cyberknife

Cyberknife radiosurgery has been introduced in the 1990s by Prof. John R. Adler, MD [14] as a therapeutic modality in the treatment of intracranial lesions. It is a light-weighted LINAC-based radiosurgery system that uses image-guided, noninvasive fixation.

Peri-/retrobulbar anesthesia is performed for treatment planning and delivery in patients with uveal melanoma. So far, results are promising, albeit with a short follow-up and a low number of patients. Muacevic et al. [15] reported results of single-session treatments with 18–22 Gy at the 70% isodose. A good tumor response and no local recurrences and secondary neovascular glaucomas were observed in 20 patients with a median follow-up of 13 months. In another study by Zorlu et al. [16], a total dose of 60 Gy was prescribed in 3 daily fractions of 20 Gy each at the 80–85% isodose contour. So far, only preliminary results with 8 months of follow-up have been published by this group, with tumor regression observed in 3 of 5 patients and stable tumor size in the other 2 patients. No serious side effects were found in any of the patients [16]. Choi et al. [17] reported retrospectively on 6 patients treated with 3 fractions and a total dose of 36–39 Gy. Although some side effects occurred, the authors concluded that this was a safe and effective treatment.

The initial results with the cyberknife are promising and show this to be a safe, efficient and convenient treatment form for uveal melanoma. However, studies with longer follow-ups and a higher number of patients are needed to assess definitive applicability of this therapeutic modality.

Linear Accelerator

Due to easier fixation systems, stereotactic irradiation of uveal melanoma using a LINAC is usually done as a hypofractionated radiotherapy. The advantage of fractionation is the reduction of the radiation dose to the healthy structure in vicinity of the tumor possibly allowing radiation damage repair in organs at risk and reducing collateral side effects. Over time, different yet similar fixation systems have been developed for stereotactic irradiation of uveal melanoma using a LINAC. A common advantage is, that these fixation systems are noninvasive, thus increasing patient comfort, and allow reliable and reproducible immobilization of the target structures throughout the whole treatment [18–21]. Considerable efforts have also been undertaken to develop and optimize the immobilization of the eye during delineation and treatment [20, 22–24]. Georg et al. [25] investigated the different collimation systems to optimize treatment beam conformation dose gradients. They evaluated the impact of a micro-multileaf collimator on LINAC-based SRT of uveal melanoma by comparing circular arc with static conformal, dynamic arc, and intensity-modulated SRT and found that the conformal micro-multileaf collimator and dynamic arc SRT are the treatment options of choice for LINAC-based SRT of uveal melanoma. Another effort to improve dosimetric verification includes the introduction of a new radiochromic film (Gafchromic EBT) for multidimensional dosimetric verification for SRT treatments [26]. De Pooter [27] recently investigated methods to optimize treatment plans for SRT and found that partially blocked cones resulted in comparably improved conformity. The technique was reported to be less complicated and verification less time-consuming than the dynamic arc or intensity-modulated radiotherapy techniques.

Treatment itself is usually delivered using 6 MV photon beams [24, 28, 29]. Different fractionation schemes were used in the beginning [30–32]. Today, 4–5 fractions of 10–12 Gy each at the 80% isodose are usually prescribed, even though some centers apply a higher dose of up to 14 Gy per fraction [28, 29, 33]. The first successful results of uveal melanoma treatment with a LINAC were reported over 10 years ago [30–32]. Studies with longer follow-up times were published in recent years – so far, local tumor control rates appear promising with 90% and more after up to 10 years following treatment [28, 29, 33, 34].

Late secondary side effect rates in general are considered to be lower in these patients than in those treated with gamma knife radiosurgery: depending on follow-up duration, radiogenic retinopathy and opticopathy rates vary between 5 and 81 and between 9 and 64%; neovascular glaucoma rates range between 5 and 42%, with the higher rates occurring in melanomas located in the juxtapapillary region and in patients with a higher treatment dose than 5 fractions of 10 Gy at the 80% isodose [24, 28, 29, 34]. Based on histological findings in eyes enucleated due to secondary neovascular glaucoma after hypofractionated SRT, Fernandes et al. [35] recently postulated that neovascular glaucoma is a secondary effect of radiation induced in the posterior segment of the eye rather than a primary radiation damage to the anterior segment. Recently, a study found a clear dose relationship between the median dose to the lacrimal gland and tear production and dry eye syndrome. A median dose of 7 Gy/fraction to the lacrimal gland caused a 50% risk of low Schirmer results and a median dose of 10 Gy resulted in a 50% probability of dry eye syndrome [36].

Visual acuity usually declines gradually after teletherapy, especially when the tumor is closely located to important ocular structures such as the optic nerve or the macula [24, 28]. In tumors located more anteriorly, visual acuity changes were not significantly reduced 2 years after radiotherapy [29].

Discussion

Teletherapy is the most convenient form of treatment for uveal melanoma, especially if a brachytherapeutic approach is not possible due to unfavorable size or tumor location. Charged-particle therapy, mostly with protons, was established over 30 years ago as an effective therapy for uveal melanoma [37]. In contrast, stereotactic irradiation is still a 'young' technique in the treatment of uveal melanoma, being supported in the last decade by a number of long-term results showing favorable safety and efficacy results. Theoretically, proton beam therapy is superior to current forms of stereotactic irradiation techniques regarding dose deposition to ocular structures and to healthy tissues distal of the tumor, whereas the advantages of stereotactic irradiation over charged particles are the lack of need for surgical placement of radiopaque markers before radiotherapy and the ability to perform treatment without expensive, large-scale facilities [38, 39]. Overall, high local tumor control rates of 90% and above can be achieved with stereotactic irradiation, which is comparable to charged-particle therapy [6, 10, 11, 13, 28, 29, 33].

Secondary complications are present in stereotactic irradiation modalities to various degrees: radiation side effects such as conjunctivitis, blepharitis or dry eye syndrome are observed in some patients but can usually be treated efficiently [24]. Cataract development occurs in irradiated eyes – however, exact long-term numbers for stereotactic treatments are difficult to determine and cataract can be treated effectively by surgery. Radiation retinopathy, opticopathy and neovascular glaucoma are present with all forms of stereotactic irradiation – these complications are the most common cause of secondary vision loss; moreover, neovascular glaucoma is the most common cause of secondary eye loss [6, 11, 24, 28, 29, 33]. They appear most commonly in eyes with posteriorly located uveal melanomas. Hypofractionated SRT shows slightly more favorable results than gamma knife radiosurgery regarding these 3 late side effects, even though results are difficult to compare due to the lack of comparative studies [6, 11, 29]. New protocols with reduced gamma knife SRS doses (35–40 Gy) may also improve outcomes. For cyberknife radiosurgery/radiotherapy, conclusions can only be drawn after longer follow-ups with more patients.

A comparison of all these results with well-established forms of radiotherapy, especially proton and helium ion teletherapy, is still difficult. For example, neovascular glaucoma rates after protons vary between 3 and 35%, after charged ions between 29 and 42%, and after stereotactic photons between 5 and 35% [40]. Since these results not only reflect differences in treatment, but are also (to a large degree) influenced by the inclusion criteria, patient cohort, tumor characteristics, different lengths of follow-up, and different grading systems, it is not easy to draw final conclusions. So far, the proton results appear to be the 'gold standard' for intertreatment comparison and clinical results are to be compared with these results. Further studies will show whether further total dose reductions, fractionation schemes and even more localized treatment deliveries result in reduced secondary complication rates while maintaining good local tumor control.

To summarize, we may say that SRS and SRT are interesting future treatment options for uveal melanoma. They are non- or minimally invasive modalities, requiring no clip implantation prior to irradiation. Within the last 10 years, they have proven to be effective in terms of local tumor control. SRS and SRT are easily accessible, since gamma knife and especially LINAC facilities are frequent worldwide. As to public financing, SRS and SRT technologies offer an interesting approach in regard to cost-effectiveness.

Acknowledgment

I would like to acknowledge the team of experts and co-workers from the Departments of Ophthalmology as well as Radiation Oncology, Medical University of Vienna, Austria, for their invaluable contributions.

References

1 Chinela AB, Zambrano A, Bunge HJ, et al: Gamma knife radiosurgery in uveal melanomas; in Steiner L, et al (eds): Radiosurgery: Baseline and Trends. New York, Raven Press, 1992, pp 161–169.

2 Marchini G, Babighian S, Tomazzoli L, et al: Stereotactic radiosurgery of uveal melanomas: preliminary results with Gamma Knife treatment. Stereotact Funct Neurosurg 1995;64(Suppl 1):72–79.

3 Zehetmayer M, Menapace R, Kitz K, Ertl A: Experience with a suction fixation system for stereotactic radiosurgery of intraocular malignancies. Stereotact Funct Neurosurg 1995;64(Suppl 1):80–86.

4 Devin F, Regis J, Berros P, et al: Treatment of choroid melanoma by Gamma-Knife radiosurgery. J Fr Ophtalmol 1996;19:551–554.

5 Rand RW, Khonsary A, Brown WJ, et al: Leksell stereotactic radiosurgery in the treatment of eye melanoma. Neurol Res 1987;9:142–146.

6 Langmann G, Pendl G, Klaus M, et al: Gamma knife radiosurgery for uveal melanomas: an 8-year experience. J Neurosurg 2000;93(Suppl 3):184–188.

7 Langmann G, Pendl G, Mullner K, et al: High- compared with low-dose radiosurgery for uveal melanomas. J Neurosurg 2002;97(5 Suppl):640–643.

8 Fakiris AJ, Lo SS, Henderson MA, et al: Gamma-knife-based stereotactic radiosurgery for uveal melanoma. Stereotact Funct Neurosurg 2007;85:106–112.

9 Simonova G, Novotny J Jr, Liscak R, Pilbauer J: Leksell gamma knife treatment of uveal melanoma. J Neurosurg 2002;97(5 Suppl):635–639.

10 Mueller AJ, Talies S, Schaller UC, et al: Stereotactic radiosurgery of large uveal melanomas with the gamma-knife. Ophthalmology 2000;107:1381–1387.

11 Modorati G, Miserocchi E, Galli L, et al: Gamma knife radiosurgery for uveal melanoma: 12 years of experience. Br J Ophthalmol 2009;93:40–44.

12 Haas A, Pinter O, Papaefthymiou G, et al: Incidence of radiation retinopathy after high-dosage single-fraction gamma knife radiosurgery for choroidal melanoma. Ophthalmology 2002;109:909–913.

13 Zehetmayer M, Kitz K, Menapace R, et al: Local tumor control and morbidity after one to three fractions of stereotactic external beam irradiation for uveal melanoma. Radiother Oncol 2000;55:135–144.

14 Adler JR Jr, Chang SD, Murphy MJ, et al: The Cyberknife: a frameless robotic system for radiosurgery. Stereotact Funct Neurosurg 1997;69:124–128.

15 Muacevic A, Nentwich M, Wowra B, et al: Development of a streamlined, non-invasive robotic radiosurgery method for treatment of uveal melanoma. Technol Cancer Res Treat 2008;7:369–374.

16 Zorlu F, Selek U, Kiratli H: Initial results of fractionated CyberKnife radiosurgery for uveal melanoma. J Neurooncol 2009;94:111–117.

17 Choi SY, Kim MS, Yoo SY, et al: Feasibility of image-guided robotic radiotherapy using three fractions for uveal melanoma. Tumori 2009;95:720–725.

18 Miralbell R, Caro M, Weber DC, et al: Stereotactic radiotherapy for ocular melanoma: initial experience using closed eyes for ocular target immobilization. Technol Cancer Res Treat 2007;6:413–417.

19 Bogner J, Petersch B, Georg D, et al: A noninvasive eye fixation and computer-aided eye monitoring system for linear accelerator-based stereotactic radiotherapy of uveal melanoma. Int J Radiat Oncol Biol Phys 2003;56:1128–1136.

20 Dieckmann K, Bogner J, Georg D, et al: A linac-based stereotactic irradiation technique of uveal melanoma. Radiother Oncol 2001;61:49–56.

21 Muller K, Nowak PJ, Luyten GP, et al: A modified relocatable stereotactic frame for irradiation of eye melanoma: design and evaluation of treatment accuracy. Int J Radiat Oncol Biol Phys 2004;58:284–291.

22 Dieckmann K, Georg D, Zehetmayer M, et al: LINAC based stereotactic radiotherapy of uveal melanoma: 4 years clinical experience. Radiother Oncol 2003;67:199–206.

23 Petersch B, Bogner J, Dieckmann K, et al: Automatic real-time surveillance of eye position and gating for stereotactic radiotherapy of uveal melanoma. Med Phys 2004;31:3521–3527.

24 Dieckmann K, Georg D, Bogner J, et al: Optimizing LINAC-based stereotactic radiotherapy of uveal melanomas: 7 years' clinical experience. Int J Radiat Oncol Biol Phys 2006;66(4 Suppl):47–52.

25 Georg D, Dieckmann K, Bogner J, et al: Impact of a micromultileaf collimator on stereotactic radiotherapy of uveal melanoma. Int J Radiat Oncol Biol Phys 2003;55:881–891.

26 Sturtewagen E, Fuss M, Paelinck L, et al: Multi-dimensional dosimetric verification of stereotactic radiotherapy for uveal melanoma using radiochromic EBT film. Z Med Phys 2008;18:27–36.

27 de Pooter JA: Plan Optimization for Stereotactic Radiotherapy; thesis, Erasmus University Medical Center Rotterdam, 2008.

28 Krema H, Somani S, Sahgal A, et al: Stereotactic radiotherapy for treatment of juxtapapillary choroidal melanoma: 3-year follow-up. Br J Ophthalmol 2009;93:1172–1176.

29 Muller K, Nowak PJ, de Pan C, et al: Effectiveness of fractionated stereotactic radiotherapy for uveal melanoma. Int J Radiat Oncol Biol Phys 2005;63: 116–122.
30 Bellmann C, Fuss M, Holz FG, et al: Stereotactic radiation therapy for malignant choroidal tumors: preliminary, short-term results. Ophthalmology 2000;107:358–365.
31 Tokuuye K, Akine Y, Sumi M, et al: Fractionated stereotactic radiotherapy for choroidal melanoma. Radiother Oncol 1997;43:87–91.
32 Zehetmayer M, Dieckmann K, Kren G, et al: Fractionated stereotactic radiotherapy with linear accelerator for uveal melanoma – Preliminary Vienna results. Strahlenther Onkol 1999;175(Suppl 2): 74–75.
33 Dunavoelgyi R, Dieckmann K, Gleiss A, et al: Local tumor control, visual acuity, and survival after hypofractionated stereotactic photon radiotherapy of choroidal melanoma in 212 patients treated between 1997 and 2007. Int J Radiat Oncol Biol Phys 2010, E-pub ahead of print.
34 Somani S, Sahgal A, Krema H, et al: Stereotactic radiotherapy in the treatment of juxtapapillary choroidal melanoma: 2-year follow-up. Can J Ophthalmol 2009;44:61–65.
35 Fernandes BF, Weisbrod D, Yucel YH, et al: Neovascular glaucoma after stereotactic radiotherapy for juxtapapillary choroidal melanoma: histopathologic and dosimetric findings. Int J Radiat Oncol Biol Phys 2010, E-pub ahead of print.
36 Muller K, Nowak PJ, Naus N, et al: Lacrimal gland radiosensitivity in uveal melanoma patients. Int J Radiat Oncol Biol Phys 2009;74:497–502.
37 Gragoudas ES: Proton beam irradiation of uveal melanomas: the first 30 years. The Weisenfeld Lecture. Invest Ophthalmol Vis Sci 2006;47:4666–4673.
38 Hocht S, Stark R, Seiler F, et al: Proton or stereotactic photon irradiation for posterior uveal melanoma? A planning intercomparison. Strahlenther Onkol 2005;181:783–788.
39 Zytkovicz A, Daftari I, Phillips TL, et al: Peripheral dose in ocular treatments with CyberKnife and Gamma Knife radiosurgery compared to proton radiotherapy. Phys Med Biol 2007;52:5957–5971.
40 Lodge M, Pijls-Johannesma M, Stirk L, et al: A systematic literature review of the clinical and cost-effectiveness of hadron therapy in cancer. Radiother Oncol 2007;83:110–122.

Martin Zehetmayer, MD
Department of Ophthalmology and Optometry, Medical University of Vienna
Waehringer Guertel 18–20
AT–1090 Vienna (Austria)
Tel. +43 1 40400 2303, E-Mail martin.zehetmayer@meduniwien.ac.at

Jager MJ, Desjardins L, Kivelä T, Damato BE (eds): Current Concepts in Uveal Melanoma.
Dev Ophthalmol. Basel, Karger, 2012, vol 49, pp 66–80

Local Resection of Uveal Melanoma

Bertil E. Damato

Royal Liverpool University Hospital, Liverpool, UK

Abstract

Local resection of uveal melanoma is aimed at conserving the eye and useful vision while removing any threat of metastatic spread. The tumour can be removed en bloc through a scleral opening (i.e., 'exoresection') or in a piecemeal fashion with a vitreous cutter passed through the retina (i.e., 'endoresection'). Variations of exoresection include iridectomy, iridocyclectomy, cyclochoroidectomy, and choroidectomy. Endoresection can be performed through a retinotomy or under a large retinal flap. Both exoresection and endoresection can be undertaken as a primary procedure, or after other conservative therapy as treatment for recurrent or toxic tumour. Each can be performed in combination with some form of radiotherapy, which can precede or follow the surgical resection. Endoresection should be relatively straightforward for experienced vitreoretinal surgeons; however, exoresection is more challenging, particularly with large and posterior tumours, because of the need for hypotensive anaesthesia and other measures to control intra-operative haemorrhage. In addition to their technical complexities, exoresection and endoresection are limited by intuitive concerns regarding iatrogenic tumour dissemination. When these obstacles are overcome, local resection preserves eyes that would otherwise be inoperable and produces relatively large tumour samples, which are useful for prognostication and research and which may one day have therapeutic value.

Surgical excision of tumours is the oldest and most common form of treatment for cancer; however, local resection of uveal melanoma is not widely performed, except for small iris and ciliary body tumours. This is because of technical difficulties and because of concerns that the surgical manipulations might disseminate tumour cells from the eye to cause metastatic death. However, these obstacles have largely been overcome, at least in some centres, thanks to developments in surgical technique and in our understanding of the metastatic process.

In this chapter, I describe the technical and theoretical developments that have culminated in the current state of the art of local resection. I will rely mostly on my personal experience of more than 600 exoresections and 60 endoresections, which I have performed over the past 26 years. I hope this article will encourage readers to consider local resection when other forms of treatment are unlikely to conserve the eye and useful vision.

Iridectomy

Iridectomy has long been performed for iris tumours and in many centres it is still the first choice of treatment for such lesions; however, the coloboma causes significant morbidity [1]. The most significant advances have been the development of artificial iris implants and improvements in pupilloplasty [2]. This procedure is indicated for tumours involving up to a third of the iris and not extending to the angle. In 1994, I replaced iridectomy with proton beam radiotherapy, which avoids the problems of surgical coloboma [3]. Such radiotherapy is not restricted by the extent of the tumour and has been used successfully to treat the entire anterior segment in patients with a diffuse melanoma. Some authors use customised radioactive plaques for this purpose [4].

There is a lack of consensus as to whether iris tumours should all be biopsied for prognostic studies.

Iridocyclectomy

As with iridectomy, this is an old operation, which is indicated for tumours involving the angle or ciliary body and with circumferential spread not exceeding 3 clock hours [5]. The standard approach is to perform broad iridectomy, after dilating the pupil, and then to extend the dissection posteriorly to include the ciliary body. Several years ago, I modified my technique, constricting the pupil pre-operatively and performing the dissection in a postero-anterior or circumferential direction. This makes it possible to conserve most or all of the iris in the affected sector so that the patient is left with a peripheral iridectomy and minimal visible external evidence of the procedure (fig. 1). Some have advocated the use of the Flieringa ring, but I have always managed well enough without this device.

My indications for iridocyclectomy have changed over the years with a swing towards proton beam radiotherapy followed by a return back to surgery again. The main advantage of surgical resection over radiotherapy is that it provides abundant tissue for diagnosis and prognostication, the scope of which has increased with advances in immunohistochemistry and genetic analysis [6].

Choroidectomy

This operation has always been controversial and still is, because of its technical difficulties and because of concerns about iatrogenic tumour dissemination. It is therefore

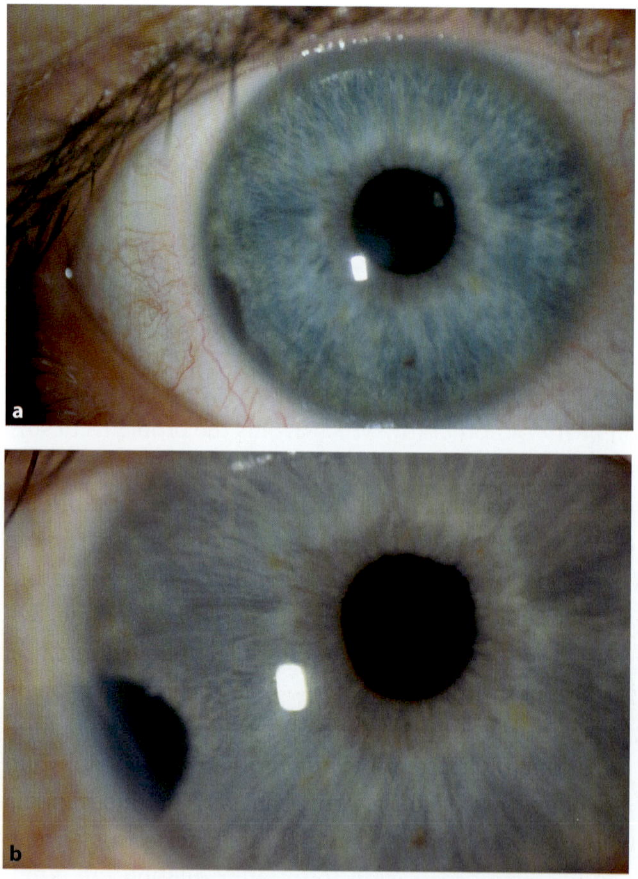

Fig. 1. Right eye of a 32-year-old man with an inferotemporal iridociliary tumour. **a** Pre-operatively with a deeply pigmented tumour, which on ultrasonography measured 1.8 mm in diameter with a thickness of 1.1 mm. **b** After cyclo-iridectomy, showing conservation of the iris sphincter. The visual acuity of 6/6 was conserved. Histology showed the tumour to be a melanoma with epithelioid cells.

performed only by a very small number of surgeons around the world. I started performing this operation in 1984, when I was training with Wallace Foulds in Glasgow. His technique was similar to that of Stallard but his indications were less exclusive so that he performed local resection as a primary procedure irrespective of the status of the fellow eye [7, 8].

Previous Technique

When I was taught choroidectomy, the technique was as follows. Any extra-ocular muscles in the operative field were disinserted. The tumour was localised by transpupillary transillumination. A rectangular, lamellar scleral flap was then prepared, hinged posteriorly. A sclerotomy was made with a feather blade and extended around the tumour with scissors, first anteriorly, then laterally and finally posteriorly. This incision was placed 1 mm within the external scleral incisions so as to create a stepped wound edge, which facilitated closure of the eye. A choroidotomy was then made with a feather blade and extended with blunt-tipped, corneoscleral, spring scissors. If the tumour involved the ciliary body, iridocyclochoroidectomy was performed. If the tumour was adherent to the retina, the two tissues were separated by blunt dissection, using spring scissors. The sclera was closed with 8-0, interrupted virgin silk sutures. Gas tamponade was administered routinely. Brachytherapy was administered only if local tumour recurrence occurred, using a 20-mm ruthenium plaque. If a retinal tear occurred, the patient was kept under close observation and vitreoretinal surgery was performed only if retinal detachment occurred. At that time, Peyman et al. [9] were advocating a full-thickness scleral flap and Shields and Shields [10] preferred a circular scleral flap. Hypotensive anaesthesia was used to lower the systolic blood pressure to approximately 40 mm Hg.

In the 1980s, choroidectomy and cyclochoroidectomy rarely took less than 3 h to complete. Intra-operative and postoperative complications were frequent. The most common accidents were: incorrect location of the scleral window; bulging of the retina through the scleral window so that large retinal tears occurred when dissecting the posterior choroid; inadvertent retinal incisions when incising the deep sclera or uvea; retinal dialysis and detachment after cyclochoroidectomy, despite plombage and internal tamponade; retinal breaks occurring when separating the tumour from adherent retina; severe haemorrhage despite hypotensive anaesthesia; subretinal haematoma developing during flap closure, and breakage of virgin silk sutures. Postoperatively, local tumour recurrence and retinal detachment were more common. In some patients, the detection of the recurrent tumour was delayed so that extra-ocular spread occurred, with multiple tumour nodules developing around the eye and orbit. Retinal detachment surgery often failed, not only because the surgical techniques were less advanced than they are today, but also because of severe proliferative vitreoretinopathy, which developed rapidly if treatment was delayed. For these reasons, many patients required secondary enucleation. Diplopia was another common problem, which was difficult to manage because of the thin sclera and the presence of extensive adhesions. Around 1986, Foulds introduced ocular decompression as a means of preventing retinal bulging during choroidectomy and this made the tumour resection easier and safer; however, he advocated total vitrectomy, using the three-port system, which I found difficult and which was sometimes complicated by lens touch and other problems. Around the same time, Foulds and I introduced adjunctive brachytherapy, delivering a dose of 500 Gy to the sclera using a 20-mm ruthenium applicator; however, this was abandoned because of complications, which included optic neuropathy, maculopathy and scleral wound dehiscence.

Current Technique

Over the years, I have developed several modifications, which might seem trivial but which have made the operation easier and safer. The operating time has become shorter so that it is not uncommon for large procedures to be completed in less than 100 min. I will briefly describe the most important developments in my resection technique.

The muscles in the operative field are disinserted, but only after measuring the knot-to-limbus distances, so that these measurements are the same when the muscles are re-attached. Oblique muscles are no longer re-attached but left loose. Thanks to these minor modifications, postoperative diplopia is now very rare.

The transillumination is now performed with a right-angled, 21-gauge fibre-optic light, which can be shone into the eye not only through the pupil but also transsclerally from the other side of the eye, thereby avoiding the penumbra, which with thick tumours used to exaggerate the apparent posterior tumour extent. When dissecting the lamellar flap posteriorly, a mark is made on the shaft of the Desmarres scarifier so that when this mark is aligned with the anterior tumour margin, the tip of the blade is located at the posterior tumour edge. This simple procedure ensures that the lamellar scleral dissection extends far enough posteriorly.

Over the years, I have changed the shape of the scleral flap so that it is now polyhedral, with the radial incisions lateral to the tumour diverging posteriorly. The polyhedral shape facilitates wound apposition during closure. The diverging radial incisions enhance access to the posterior part of the tumour so that they can be shorter, requiring fewer sutures and saving time.

Only a limited vitrectomy is now performed and this is done through a single sclerotomy, using an O'Malley contact lens and light from the operating microscope, indenting the eye instead of using an infusion. This minimalist approach has proved quicker, easier and safer.

When the anterior edge of the flap is located along the limbus, as is necessary for cyclochoroidectomy, this edge is also stepped, like the other margins, thereby preventing wound leakage and hypotony. This is done by splitting the deep sclera into two layers and performing the lamellar dissection in an anterior direction into cornea.

The problem of persistent intra-operative haemorrhage despite hypotensive anaesthesia has diminished considerably since I started cauterising the short posterior ciliary arteries next to the optic nerve in the quadrant of the tumour. The first time I did this was with a patient (the sister of an eminent ophthalmologist) whose cerebral function deteriorated when the blood pressure was lowered. Closure of these blood vessels allowed me to complete the procedure without hypotension and the eye was saved with visual acuity of 6/12 (albeit with unilateral normal-tension glaucoma).

Instead of incising the deep sclera with the blade held perpendicular to the flat scleral surface, I now pinch the sclera with toothed microforceps and shave the scleral fold with the blade held tangential to the sclera. Furthermore, I abandoned

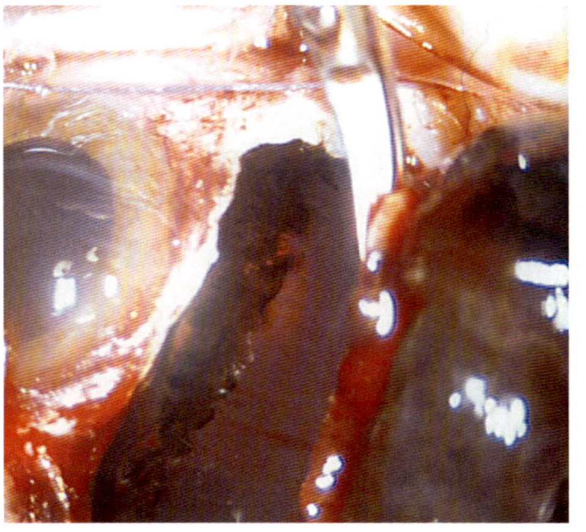

Fig. 2. Intra-operative photograph during exoresection of an anterior choroidal melanoma, showing conservation of the ciliary epithelium in the region of the pars plana.

the incision technique for making the choroidotomy, instead grasping the uvea with two pairs of notched microforceps and pulling these apart until a rip occurred. These minor adjustments have eliminated inadvertent retinal tears during sclerotomy and choroidotomy.

Around 1999, it occurred to me that retinal dialysis and detachment might not occur if it was possible to conserve the ciliary epithelium over the pars plana. However, this seemed impossible as the ciliary epithelium is only two cells thick. I eventually solved this problem by abandoning iridocyclochoroidectomy and doing the operation back-to-front, that is, by performing choroidocyclectomy. This is done by making the choroidotomy posterior to the ora serrata, where the retina is relatively strong, then separating the pars plana epithelium from the uvea by blunt dissection, and snipping the uvea without allowing the scissors to touch the epithelium, lifting the uvea with microforceps. It has proved surprisingly easy to conserve the pars plana epithelium, and this has eliminated the common and severe complication of retinal dialysis (fig. 2).

The next major challenge was to avoid retinal breaks when separating tumour apex from adherent retina. This problem is more common with thick tumours, especially those that have ruptured Bruch's membrane [11]. Counterintuitively, the key to conserving an intact retina is to use a Bard-Parker scalpel instead of attempting blunt dissection with closed scissors. I discovered this serendipitously when I started using a blade with the intention of 'top-slicing' the adherent tumour apex, leaving it on the retina and subsequently treating it with brachytherapy. I found that on scraping the tumour surface a few millimetres away from the retina, invisible strands were divided so that the retina peeled away from the tumour. This has proved to be a major advance, greatly reducing the incidence of rhegmatogenous retinal detachment.

The common problem of subretinal haematoma developing during scleral suturing has been resolved (to a large extent but not completely) by wedging two sponge cells hard against the eye so that they compress the uveal blood vessels posterior to the coloboma. Simultaneously, the intra-ocular pressure is increased by gently pulling on the traction sutures until the retina bulges slightly through the scleral window. It might seem a good idea to apply diathermy to the uvea, but this is associated with a risk of causing retinal burns and tears and is to be avoided.

When a retinal break occurs, this is treated by vitreoretinal surgery, which is done as soon as the scleral flap is closed. Such surgery is performed by a vitreoretinal colleague, who performs full vitrectomy, evacuates any subretinal haemorrhage, flattens the eye with heavy liquid, applies endolaser for retinopexy, and fills the eye with silicone. Such surgery is almost always successful in preventing retinal detachment. Gas tamponade, external plombage and the old 'wait-and-see' policy have all been abandoned.

The incidence of local tumour recurrence has diminished greatly since the reintroduction of adjunctive brachytherapy using a 25-mm ruthenium plaque. The 20-mm applicator was not large enough, which in retrospect is not surprising considering the large basal dimensions of tumours treated by exoresection. Recurrent tumour now occurs only in areas that are not treated with the plaque, for example, if the tumour and hence the coloboma has an irregular shape curving around the optic disc. The radiotherapy does not prevent satellite recurrences in distant parts of the uvea, which are fortunately very rare [12]. The adjunctive brachytherapy is not without risks, which include: delayed wound healing and dehiscence; optic neuropathy; cyclodialysis and hypotony, and scleral necrosis. These can be prevented by: using non-absorbable sutures for scleral closure; using a perforated template to facilitate proper plaque location and performing ophthalmoscopy while transilluminating through the perforations; delaying brachytherapy by 1 month if cyclectomy is performed, and ensuring good conjunctival cover to avoid exposure of irradiated sclera.

The adjunctive brachytherapy not only prevents local tumour recurrence but also decreases postoperative morbidity by reducing the need for wide safety margins and therefore allowing more conservation of the iris and ciliary body.

Results

Several published articles report the results of choroidectomy, but these are now out of date [9, 10, 13–15]. I recently reviewed the results of 112 primary exoresections for uveal melanomas involving choroid, which were performed between 2000 and 2010 (fig. 3). An article is in preparation and data are still being collected. Briefly, the patients had a median age of 48 years. The tumours had a median diameter of 15.3 mm and a median thickness of 8.5 mm with extension anterior to the ora serrata in 41% and to within 3 mm of the disc or fovea in 16%. Eighty-eight percent of eyes were conserved with retention of vision of 6/60 or better in 58% and of 6/12 or

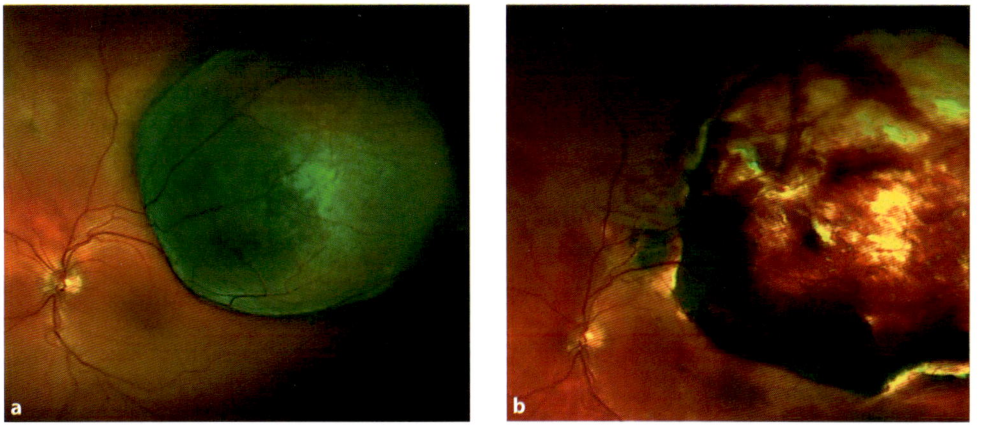

Fig. 3. Left fundus of a 51-year-old man with a superotemporal choroidal melanoma of the left eye, the tumour measuring 17.1 mm in basal diameter with a thickness of 7.4 mm. **a** Pre-operative photograph showing the dome-shaped tumour extending close to the fovea. The visual acuity was 6/18. **b** Fundus appearance on the second postoperative day, when the visual acuity was still 6/18.

better in 30%. The main reasons for enucleation were tumour recurrence (5 patients), abandoned procedure (3 patients), retinal detachment (3 patients) and 'other' in 2 patients. Local tumour recurrence and retinal detachment occurred in 11 eyes, some of which received no brachytherapy or were treated with a 20-mm applicator, which, as mentioned before, is now known to be inadequate. Rhegmatogenous retinal detachment developed in 14 eyes with about 30% of these conserving vision of 6/12 or better. Nineteen patients died, 16 of these as a result of metastatic disease. This study included some patients whom I considered unsuitable for local resection, because of advanced tumour or because hypotensive anaesthesia was contraindicated, but whom I felt obliged to treat conservatively because they had refused enucleation. One such patient had local resection without hypotension because she had the thallasaemia trait and the other had a large tumour with total retinal detachment touching the lens and with an intra-ocular pressure of 44 mm Hg. Good results were obtained in both patients. Serious anaesthetic complications have been rare, with 1 patient in the 1980s dying of pulmonary embolism (probably unrelated to the anaesthesia) and another, in 2010, developing a postoperative, cardiac arrest, from which he made a full recovery.

Current Indications

I currently perform primary exoresection for patients whose tumour is too thick for ruthenium plaque radiotherapy (i.e., with a thickness exceeding 5–6 mm) and who are likely to develop canalicular damage or neovascular glaucoma after proton beam radiotherapy. In the 1990s, I shifted from resection to proton beam radiotherapy but then went back to local resection once I saw that radiation-induced complications were more common than I had expected. When anti-angiogenic agents were introduced I once again became more optimistic about proton beam radiotherapy but was disappointed with the results of anti-angiogenic agents. These experiences, together with concurrent improvements in surgical resection techniques, have swung me back to exoresection again.

Toxic Tumour Syndrome

Most secondary exoresections are undertaken as a treatment for what I have termed the 'toxic tumour syndrome'. This condition develops when an irradiated intra-ocular tumour becomes ischaemic and exudative as a result of intratumoural, radiation-induced vasculopathy, leading to effects such as macular oedema and exudates, serous retinal detachment, uveitis, rubeosis iridis and neovascular glaucoma. I have previously reported the case of a 75-year-old man with a nasal, equatorial choroidal melanoma, which had a basal diameter of 13.0 mm and a thickness of 9.3 mm, reducing the visual acuity to 6/36 [16]. The fellow eye was amblyopic with a vision of 6/36. The patient had a history of cerebrovascular disease, ischaemic heart disease and cardiac arrhythmia and was receiving anticoagulant therapy. I therefore administered proton beam radiotherapy, but this was complicated by severe retinal detachment and neovascular glaucoma with an intra-ocular pressure of 50 mm Hg. Because of the amblyopia in the fellow eye, I felt obliged to attempt exoresection, which was performed with limited hypotension, without complications. Postoperatively, the retina was flat by the first postoperative day and the rubeosis regressed completely within a few weeks so that after several months the intra-ocular pressure remained normal after discontinuing the topical antiglaucomatous medications. To date, I have performed secondary exoresection for toxic tumour syndrome in less than 20 patients, with encouraging results, which I hope to report soon. Local resection can be effective for toxic tumour syndrome occurring with irradiated adenocarcinoma, and presumably other types of tumour [17].

Endoresection

'Endoresection' is a term I coined several years ago to refer to piecemeal removal of an intra-ocular tumour with a vitreous cutter [18]. Peyman has used the phrase 'ab interno resection' for the same kind of procedure [19].

My technique has been described previously (fig. 4). Briefly, after total vitrectomy, the vitreous cutter is passed through the retina at the apex of the tumour, which is removed together with a surround of apparently normal choroid. Fluid-air exchange

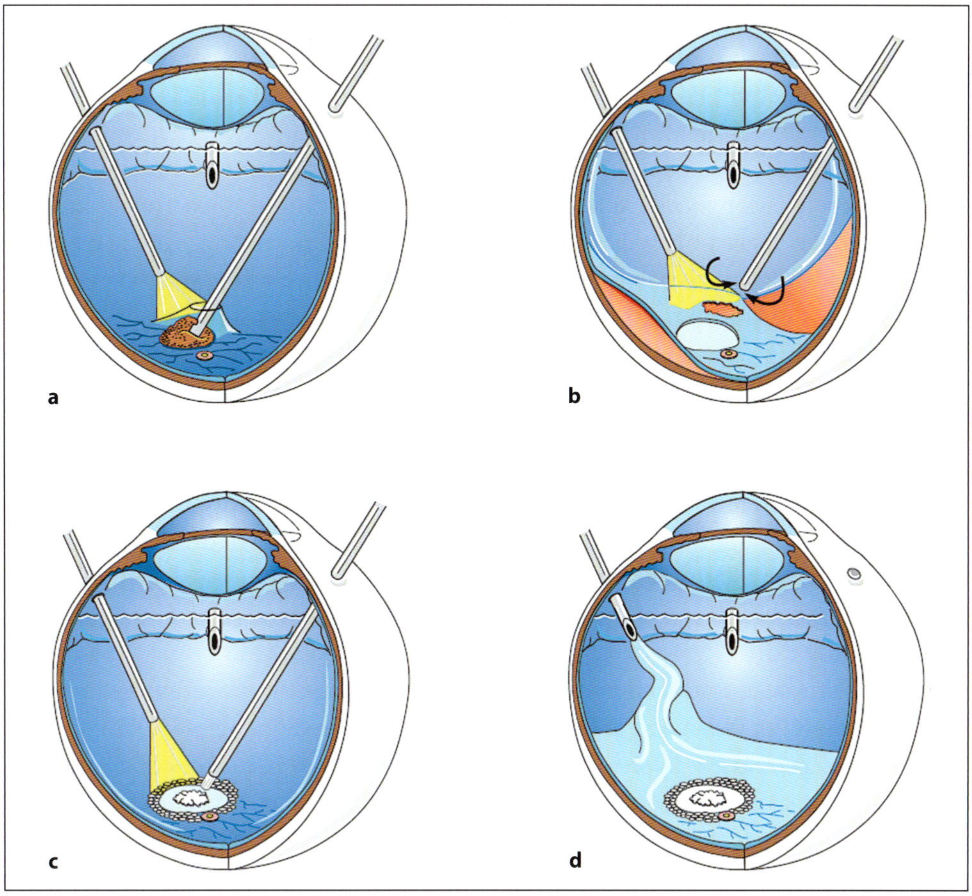

Fig. 4. Technique of endoresection. **a** After total vitrectomy, the tumour is removed through a small retinotomy. **b** Fluid-air exchange is performed, to flatten the retina. **c** Endolaser photocoagulation is applied to destroy any residual tumour and to achieve retinopexy. **d** The eye is filled with silicone to prevent postoperative haemorrhage and retinal detachment. This oil is removed after 12 weeks. The procedure is easier if performed bimanually, using a chandelier for illumination.

is performed to flatten the retina so that retinopexy can be performed, which is done with endolaser photocoagulation. Endolaser is also applied to the bed of the surgical coloboma to treat any intrascleral tumour remnants, which can also be treated externally with a right-angled applicator passed retro-ocularly. The air is replaced with silicone, which is removed after 12 weeks. Cryotherapy is applied to the sclerotomies, using a double freeze-thaw technique, in case of seeding. If histology indicates high-grade malignancy, then adjunctive brachytherapy is delivered using a ruthenium applicator.

Primary endoresection is highly controversial because of concerns that the surgical manipulations might disseminate tumour cells around the globe and systemically [20]. Such objections are based on mechanistic concepts of metastatic spread.

However, insights from genetic studies suggest that metastasis is a biological process, with almost all monosomy 3/class 2 melanomas proving fatal, irrespective of any surgical intervention [21–23]. As with other forms of treatment, intrascleral and marginal tumour remnants can give rise to extra-ocular and marginal recurrences; however, in the absence of an intact retinal barrier, such recurrent tumours can seed into the vitreous cavity to become inoperable [24, 25]. In more than 100 endoresections, there has been only 1 patient with subconjunctival seeding, which seems to have been adequately controlled by cryotherapy. Mortality rates have been similar to other forms of therapy, although formal comparisons have yet to be undertaken. The most common complications are related to the vitrectomy procedure and these include entry site tears, cataract, and, rarely, transient, acute glaucoma in the early postoperative period. Rhegmatogenous retinal detachment has become rare since the introduction of peripheral retinal inspection and cryotherapy during the endoresection procedure. A few patients have developed choroidal neovascular membranes arising from the margin of the coloboma. The visual outcome depends on the location of the tumour with conservation of visual acuity being best when the tumour is nasal to the optic disc. Others have reported encouraging results, even with large tumours [26].

Some advocate endoresection under a large retinal flap, but my impression is that this is not useful unless macular translocation is performed.

I consider primary endoresection only if it offers the best chances of conserving useful vision (e.g. if radiotherapy is likely to cause optic neuropathy) and if the patient accepts the controversial nature of this procedure. Preservation of central vision is less likely if a juxtapapillary tumour involves the temporal disc margin or if it extends close to the fovea. Without adjunctive radiotherapy, the risk of local tumour recurrence is greater if the tumour has diffuse margins or if its basal diameter exceeds 10–11 mm.

Secondary endoresection is an effective treatment for the toxic tumour syndrome and is performed if the tumour is too large for transpupillary thermotherapy and too small and posterior for transscleral local resection. Endoresection is also useful as a treatment for local tumour recurrence after other forms of treatment.

Some authors advocate neo-adjuvant proton beam or stereotactic radiotherapy before endoresection [27, 28]; however, after endoresection alone local tumour recurrence is rare, which suggests that if radiotherapy were to be administered to all patients, the large majority would be subjected to this treatment unnecessarily. In Liverpool, local recurrence is probably infrequent because endoresection is reserved for tumours not exceeding 10–11 mm in basal diameter and such small tumours are less likely to be of high-grade malignancy and to have long, invisible, diffuse, marginal extensions. The scope of neo-adjuvant radiotherapy is likely to be greater with large, highly malignant melanomas, especially if these are located far enough from the optic nerve and fovea to avoid collateral damage to these structures.

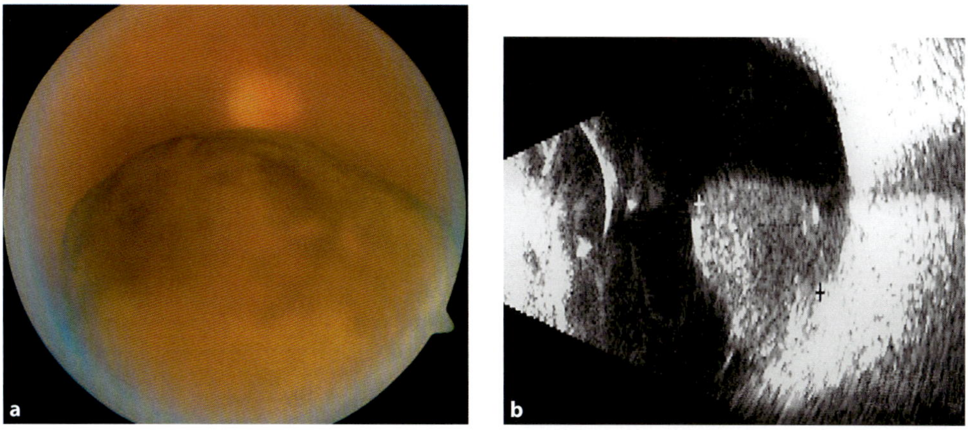

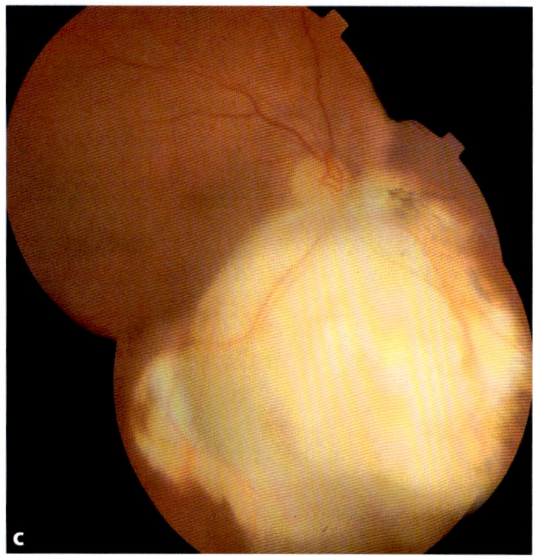

Fig. 5. Right fundus of a 65-year-old man with an inferonasal choroidal melanoma. **a** Pre-operative photograph. **b** Ultrasound scan showing the tumour extending close to the optic disc and having a basal diameter of 10.4 mm with a thickness of 9.6 mm. **c** Fundus appearance after endoresection, which was performed in 1999 by the author. The patient was alive more than 10 years postoperatively and after cataract surgery maintained a visual acuity of 6/6.

Discussion

Since the theme of this book series is 'Developments in ophthalmology', I will focus on the developmental aspects of local resection of uveal melanoma.

As with many other surgical procedures, progress has come about mostly by the mental process of reflecting on complications and thinking about how these might be avoided. Examples of innovations that have occurred in this fashion include sclerotomy and choroidotomy techniques that have greatly reduced the incidence of retinal tears. Some developments occurred only after years of thought and one such example is the prevention of retinal dialysis and detachment by conservation of the ciliary epithelium over the pars plana. Other improvements, such as cauterisation of the short posterior ciliary arteries, occurred when a crisis arose intra-operatively. Some procedures, such as adjunctive brachytherapy, are based on evidence gained from long-term outcome analyses. Although the instruments I use for exoresection are the same as those I deployed when I first started, more than 25 years ago, the equipment required for endoresection has improved greatly, making this operation safer and easier. Examples of such technical advances include the air pump and the chandelier, which has made bimanual surgery possible.

During the years in which I have been performing local resection, my attitudes regarding the impact of my surgery on survival have changed several times, in keeping with the prevailing beliefs. When I started in 1984, many surgeons were performing enucleation as an urgent procedure to prevent metastasis and a large number followed Zimmerman's hypothesis, using a no-touch technique or administering pre-enucleation radiotherapy, to prevent iatrogenic tumour dissemination [29]. There were concerns, therefore, that the surgical manipulations required by local resection were dangerous. However, non-randomised comparative studies showed no significant survival difference between enucleation and exoresection [30]. Insights from genetic studies subsequently suggested that ocular treatment of uveal melanoma did not influence survival, which was already determined by the time the patient was treated, because lethal and non-lethal melanomas were distinct from their inception [31, 32]. Recent evidence suggests that some disomy 3 melanomas can undergo late malignant transformation to monosomy 3 [33]. If such transformation were to occur in a recurrent tumour after local resection, then some patients could die of metastatic disease that might have been preventable by more radical treatment. Bechrakis et al. [34] have shown that recurrent melanoma after exoresection may show higher-grade malignancy than the resected tumour, but whether this represents malignant transformation or overgrowth of indolent, spindle melanoma cells by epithelioid cells is not known. Recurrent uveal melanoma is known to be associated with increased mortality [35]; however, there is uncertainty as to whether the recurrence is the source of the metastasis or merely an indicator of increased malignancy. Randomised studies would be necessary to address this question. The Collaborative Ocular Melanoma Study reported no difference in survival between brachytherapy and enucleation; however, the number of patients with recurrent tumour was too small for statistical significance to be demonstrable [36]. In view of the rarity of local tumour recurrence after local resection of disomy 3 melanoma, the chances of malignant transformation and metastatic spread of residual tumour must be small.

It is difficult to predict how the various types of local resection will continue to develop in future years. The greatest obstacle preventing more widespread exoresection is the need for profound hypotensive anaesthesia, which is controversial and therefore administered only by few anaesthetists. The rarity of complications suggests that hypotensive anaesthesia is reasonably safe if undertaken by a skilled anaesthetist equipped with adequate facilities for cerebral and cardiac monitoring. In any case, there is scope for improved methods of haemostasis so that the need of systemic hypotension is reduced. Experience with cauterisation of the short posterior ciliary arteries suggests that endoscopic placement of temporary microclips may perhaps be useful. Another hurdle to overcome is the steep learning curve, as few surgeons will have the opportunity of serving an apprenticeship with a surgeon experienced in exoresection and of undertaking the procedure often enough to maintain and develop surgical skills. Advances in videoconferencing, telemedicine and robotics may prove useful in this regard. My personal experience has shown that many patients are prepared to travel to a distant centre for exoresection if only they are informed of this by their local ophthalmologist. It would be ideal if 4 or 5 centres in Europe, for example, could include local resection in their repertoire of therapies. Endoresection is less problematic, because it does not require hypotensive anaesthesia and because it should be within the competence of any experienced vitreoretinal surgeon.

Conclusions

Developments in surgical technique have greatly improved outcomes after local resection, largely thanks to better prevention and treatment of rhegmatogenous retinal detachment and local tumour recurrence; however, concerns regarding hypotensive anaesthesia and iatrogenic metastatic spread prevent more widespread adoption of this treatment. These fears seem exaggerated and hopefully will be dispelled by further studies so that the benefits of local resection can become available to more patients. Prospects for further development of local resection techniques would be enhanced if a working group were to be formed to share experience and results.

References

1. Arentsen JJ, Green WR: Melanoma of the iris: report of 72 cases treated surgically. Ophthalmic Surg 1975;6:23–37.
2. Sakellaris D, Goossens C, Tassignon MJ: Customized iris clip anterior chamber intraocular lenses designed for iris reconstruction. Eur J Ophthalmol 2009;19:1084–1087.
3. Damato B, Kacperek A, Chopra M, Sheen MA, Campbell IR, Errington RD: Proton beam radiotherapy of iris melanoma. Int J Radiat Oncol Biol Phys 2005;63:109–115.
4. Shields CL, Shields JA, De Potter P, Singh AD, Hernandez C, Brady LW: Treatment of non-resectable malignant iris tumours with custom designed plaque radiotherapy. Br J Ophthalmol 1995;79:306–312.
5. Damato BE, Foulds WS: Ciliary body tumours and their management. Trans Ophthalmol Soc UK 1986;105:257–264.
6. Damato EM, Damato B, Sibbring JS, Coupland SE: Ciliary body melanoma with partial deletion of chromosome 3 detected with multiplex ligation-dependent probe amplification. Graefes Arch Clin Exp Ophthalmol 2008;246:1637–1640.
7. Foulds WS: Experience of local excision of uveal melanomas. Trans Ophthalmol Soc UK 1977;97:412–415.
8. Stallard HB: Partial choroidectomy. Br J Ophthalmol 1966;50:660–662.
9. Peyman GA, Juarez CP, Diamond JG, Raichand M: Ten years experience with eye wall resection for uveal malignant melanomas. Ophthalmology 1984;91:1720–1725.
10. Shields JA, Shields CL: Surgical approach to lamellar sclerouvectomy for posterior uveal melanomas: the 1986 Schoenberg lecture. Ophthalmic Surg 1988;19:774–780.
11. Damato B, Groenewald CP, McGalliard JN, Wong D: Rhegmatogenous retinal detachment after transscleral local resection of choroidal melanoma. Ophthalmology 2002;109:2137–2143.

12 Kim JW, Damato BE, Hiscott P: Noncontiguous tumor recurrence of posterior uveal melanoma after transscleral local resection. Arch Ophthalmol 2002; 120:1659–1664.
13 Damato BE, Paul J, Foulds WS: Predictive factors of visual outcome after local resection of choroidal melanoma. Br J Ophthalmol 1993;77:616–623.
14 Damato BE, Paul J, Foulds WS: Risk factors for metastatic uveal melanoma after trans-scleral local resection. Br J Ophthalmol 1996;80:109–116.
15 Damato BE, Paul J, Foulds WS: Risk factors for residual and recurrent uveal melanoma after trans-scleral local resection. Br J Ophthalmol 1996;80:102–108.
16 Damato BE, Foulds WS: Surgical resection of choroidal melanoma; in Schachat AP, Ryan SJ (eds): Retina. St Louis, Mosby, Inc, 2006, pp 769–778.
17 Schalenbourg A, Coupland S, Kacperek A, Damato B: Iridocyclectomy for neovascular glaucoma caused by proton-beam radiotherapy of pigmented ciliary adenocarcinoma. Graefes Arch Clin Exp Ophthalmol 2008;246:1499–1501.
18 Damato B, Groenewald C, McGalliard J, Wong D: Endoresection of choroidal melanoma. Br J Ophthalmol 1998;82:213–218.
19 Peyman GA, Cohen SB: Ab interno resection of uveal melanoma. Int Ophthalmol 1986;9:29–36.
20 Damato B: Choroidal melanoma endoresection, dandelions and allegory-based medicine. Br J Ophthalmol 2008;92:1013–1014.
21 Prescher G, Bornfeld N, Hirche H, Horsthemke B, Jockel KH, Becher R: Prognostic implications of monosomy 3 in uveal melanoma. Lancet 1996;347: 1222–1225.
22 Onken MD, Worley LA, Ehlers JP, Harbour JW: Gene expression profiling in uveal melanoma reveals two molecular classes and predicts metastatic death. Cancer Res 2004;64:7205–7209.
23 Damato BE, Dopierala JA, Coupland SE: Genotypic profiling of 452 choroidal melanomas with multiplex ligation-dependent probe amplification. Clin Cancer Res 2010;16:6083–6092.
24 Damato B, Wong D, Green FD, Mackenzie JM: Intrascleral recurrence of uveal melanoma after transretinal 'endoresection'. Br J Ophthalmol 2001; 85:114–115.
25 Hadden PW, Hiscott PS, Damato BE: Histopathology of eyes enucleated after endoresection of choroidal melanoma. Ophthalmology 2004;111:154–160.
26 Garcia-Arumi J, Zapata MA, Balaguer O, Fonollosa A, Boixadera A, Martinez-Castillo V: Endoresection in high posterior choroidal melanomas: long-term outcome. Br J Ophthalmol 2008;92:1040–1045.
27 Schilling H, Bornfeld N, Talies S, et al: Endoresection of large uveal melanomas after pretreatment by single-dose stereotactic convergence irradiation with the leksell gamma knife – First experience on 46 cases. Klin Monatsbl Augenheilkd 2006;223:513–520.
28 Bechrakis NE, Hocht S, Martus P, Kreusel KM, Heese J, Foerster MH: Endoresection following proton beam irradiation of large uveal melanomas. Ophthalmologe 2004;101:370–376.
29 Zimmerman LE, McLean IW, Foster WD: Does enucleation of the eye containing a malignant melanoma prevent or accelerate the dissemination of tumour cells. Br J Ophthalmol 1978;62:420–425.
30 Foulds WS, Damato BE, Burton RL: Local resection versus enucleation in the management of choroidal melanoma. Eye 1987;1:676–679.
31 Parrella P, Sidransky D, Merbs SL: Allelotype of posterior uveal melanoma: implications for a bifurcated tumor progression pathway. Cancer Res 1999;59: 3032–3037.
32 Tschentscher F, Husing J, Holter T, et al: Tumor classification based on gene expression profiling shows that uveal melanomas with and without monosomy 3 represent two distinct entities. Cancer Res 2003;63: 2578–2584.
33 Callejo SA, Dopierala J, Coupland SE, Damato B: Sudden growth of a choroidal melanoma and multiplex ligation-dependent probe amplification findings suggesting late transformation to monosomy 3 type. Arch Ophthalmol 2011;129:958–960.
34 Bechrakis NE, Sehu KW, Lee WR, Damato BE, Foerster MH: Transformation of cell type in uveal melanomas: a quantitative histologic analysis. Arch Ophthalmol 2000;118:1406–1412.
35 Gragoudas ES, Lane AM, Munzenrider J, Egan KM, Li W: Long-term risk of local failure after proton therapy for choroidal/ciliary body melanoma. Trans Am Ophthalmol Soc 2002;100:43–48.
36 Damato B: Legacy of the collaborative ocular melanoma study. Arch Ophthalmol 2007;125:966–968.

Bertil E. Damato, MD, PhD, FRCOphth
Ocular Oncology Service, Royal Liverpool University Hospital
Prescot Street
Liverpool L7 8XP (UK)
Tel. +44 151 706 3973, E-Mail bertil@damato.co.uk

Jager MJ, Desjardins L, Kivelä T, Damato BE (eds): Current Concepts in Uveal Melanoma.
Dev Ophthalmol. Basel, Karger, 2012, vol 49, pp 81–95

Biopsies in Uveal Melanoma

Edoardo Midena[a,b] · Raffaele Parrozzani[b]

[a]Department of Ophthalmology, University of Padova, Padova, and [b]Fondazione GB Bietti per l'Oftalmologia, IRCCS, Rome, Italy

Abstract

The ability to obtain the proper clinical diagnosis in cases of suspected intraocular tumors has greatly advanced during the past 50 years. The clinical characteristics of most intraocular tumors (size, shape, color, and texture) are detectable by skilled ophthalmoscopic examination and, with the use of adjunctive techniques (mainly ultrasonography), the proper diagnosis can be reached without invasive procedures. Notwithstanding, some intraocular tumors need to be biopsied to obtain a reliable diagnosis. In the cytogenetic era, intraocular tumor management is changing, and tumor-sampling procedures are becoming the main prognostic (and theoretically also diagnostic) tools for uveal melanoma. In spite of the widespread use of biopsies in general surgical practice, in ophthalmic oncology, indications and contraindications for biopsy continue to be under debate. The purpose of this paper is to critically evaluate the role of uveal melanoma biopsy in current clinical practice.

Ophthalmic oncologists are among the few, if not the only, cancer-treating physicians who do not routinely use cyto- or histologic confirmation before treating a clinically diagnosed (or suspected) malignancy [1]. There are at least two reasons: a widely accepted high accuracy in noninvasive diagnosis of posterior uveal melanoma by expert clinicians, and the claimed risk of tumor diffusion secondary to diagnostic invasive approaches [2, 3]. The former has never been convincingly proven for any choroidal pigmented lesion with a thickness of less than 3 mm and a largest basal diameter under 10 mm [4]. Considering the latter, there is no evidence of an increased risk of uveal melanoma local diffusion following a correctly performed fine needle aspiration biopsy (FNAB) [5, 6].

Moreover, uveal melanoma shows some peculiarities compared to the majority of solid cancers: (1) disease-related mortality has not changed during the last 80 years, despite the introduction of new conservative therapeutic approaches [7]; (2) the current management of small choroidal indeterminate pigmented lesions (encompassing atypical nevi as well as 'small melanomas' in preclinical phases) is periodical observation until growth, whereas general oncology practice considers an earlier diagnosis (and treatment) as a first and mandatory step to improve patient survival [4, 8]; (3) the cytogenetic profile of posterior uveal melanoma is currently considered the most important prognostic factor for metastatic disease, raising relevant questions about the biology and the natural history of these lesions [5, 7]. These considerations suggest critically evaluating the role of uveal melanoma biopsy in current clinical practice, the extent to which this practice is supported by evidence from previous reports and defining the main indications and contraindications of different sampling techniques.

Diagnostic Intraocular Tumor Biopsy: Indications

The purpose of a diagnostic intraocular tumor biopsy is to confirm or rule out the clinical suspicion of a malignant tumor where noninvasive techniques fail [1]. General indications for diagnostic intraocular tumor biopsy include [1, 9]:

1 substantial probability that the suspected lesion is a malignant neoplasm;
2 substantial probability that the pathologic evaluation (including any laboratory analysis) will result in a definitive pathologic diagnosis;
3 different management options for the lesion under investigation.

In addition, specific indications for intraocular tumor biopsy include:

1 major diagnostic uncertainty about the pathologic nature of the intraocular mass, provided that the individual expressing the uncertainty is highly experienced in the ophthalmic oncology field;
2 suspected metastatic carcinoma to the eye in a patient without a prior history of a nonophthalmic malignancy or pathologically confirmed metastases in other tissues or organs;
3 request for biopsy by an informed patient who refuses recommended treatment in the absence of pathologic confirmation of the clinical diagnosis.

Therapeutic dilemmas mainly occur in clinical cases requiring treatment, in which noninvasive investigation has failed to establish a proper diagnosis. Major discrepancies between different noninvasive tests performed by expert clinicians should be considered as an indication for diagnostic tumor biopsy, both in globe-saving treatments and prior to enucleation [1, 10].

Intraocular Tumor Biopsy: Contraindications

An absolute contraindication to intraocular tumor biopsy probably does not exist [1, 5, 9]. Nevertheless, the ophthalmic oncologist needs to plan a biopsy approach (1) that will probably yield a representative and sufficient sample of the lesion for definitive diagnosis, and (2) without adding local or systemic morbidity [9]. However, it is mainly relevant to stress that a biopsy must not be used as a shortcut to a diagnosis: a proper and complete workup must be performed before considering any invasive technique [11].

Special care should be taken to plan intraocular biopsy in childhood intraocular tumors [12, 13]. Because of the high risk of retinoblastoma dissemination, it is mandatory to use mini-invasive approaches (FNAB only) in children, performed by expert clinicians, in very selected cases [11–13].

Intraocular Tumor Biopsy: Techniques

There are several different intraocular tumor biopsy techniques, including aqueous or vitreous tap, FNAB performed transsclerally or transvitreally, vitrectomy approaches, endoresection and transscleral resection [11]. All of these procedures are potentially diagnostic, with different complication rates and side effects.

Anterior Segment Tumors

Aqueous Tap

The aqueous tap is mainly used in the diagnosis of anterior uveal lymphoma, with the additional use of cell marker and cytokine studies [14]. Moreover, selected iris melanomas, mainly the diffuse type, and iris metastases, both characterized by aqueous seeding, are amenable to be sampled by this procedure [14]. Woog et al. [14] reported a 46-year-old woman with a history of breast carcinoma and no known metastatic disease who presented with iridocyclitis and secondary glaucoma. Cytological examination of the aqueous revealed adenocarcinoma. Char et al. [15] reported a small series of histologically confirmed iris ring melanomas diagnosed by aqueous tap. Unfortunately, cytological diagnosis of aqueous material is often complex because malignant characteristics of sampled cells (and their viability) can be difficult to be interpreted by the pathologist [15]. Although false-negative and false-positive results can occur, their incidence is low in the hands of a skillful cytopathologist [1]. Therefore, this technique should be considered as the first and least-invasive approach for selected iris lesions, with visible aqueous seeding.

Iris Fine Needle Aspiration Biopsy

Iris FNAB involves proper instrumentation, planning of the tumor approach, handling of harvested cells, and preparation and interpretation of cytological specimens [1, 16]. It is important to realize that only a limited number of cells may be obtained through aspiration. The surgical approach depends on location and tumor size. A clear cornea entry is performed, approximately 90° from the tumor. The needle should be inserted through the cornea at an approximate 20–30° angle to the iris and, when inside the anterior chamber, the needle should be parallel to the iris [16]. Most often, the preferred entry site is either temporal or inferior. A specific needling procedure (gently moving the needle back and forth into the lesion while always maintaining the vacuum) is recommended [1, 16]. Shields et al. [16] reported a diagnostic yield of 99% in 100 consecutive iris lesions sampled by FNAB (performing 1 (74%), 2 (24%) or 3 (2%) FNAB passes). These authors concluded that FNAB appears as a safe and useful diagnostic technique, providing adequate cell sampling for cytological interpretation in nearly all cases. Postoperative hyphema can occur, but generally resolves spontaneously. The major challenge in iris FNAB occurs when the specimens are paucicellular and decisions should be based on a small number of cells. As well as with the aqueous tap technique, the experience and skill of the cytopathologist are equally critical in all FNAB-based diagnoses [15, 16].

Iris Biopsy Using Vitreous Cutter

Bechrakis et al. [17] reported 11 cases of iris tumor biopsies performed by a vitreous cutter using a 2-port clear cornea approach. A 21-gauge infusion was inserted into the anterior chamber and intraocular pressure was elevated to 70 mm Hg. A 20-gauge vitreous cutter was then inserted through the second limbal incision and placed on the tumor surface in such a way that its opening was occluded by tumor tissue. With a high aspiration setting (400 mm Hg) and low cutting frequency (80/min), one single bite was obtained from the tumor surface. Tumor sampling was diagnostic in all cases. Although this technique appears to be a safe and effective method, Bechrakis et al. [17] do not recommend its use in a routine clinical setting due to its costs.

Iris Surgical Biopsies

Iris biopsy with a punch technique was shown by Pe'er et al. [18] to yield adequate tissue specimens in 2 cases. A trabeculectomy punch was inserted into a viscoelastic-filled anterior chamber. Expensive equipment, such as a vitrectomy set, was not required. This quick and safe technique was useful in a case of unsuccessful FNAB sampling. Handling of harvested tissue is simple, because the punch can be opened over a dry sponge and inspected directly instead of being received in a large volume of liquid [18].

Surgical iridectomy through a corneal or limbal incision can also provide adequate diagnostic material in most of reported cases [19]. Nevertheless, this technique is a surgical invasive procedure, with more side effects [19]. Surgical iridectomy is mainly used as excisional biopsy to remove all neoplastic tissue.

Posterior Segment Tumors

Fine Needle Aspiration Biopsy

A tumor wider than 10 mm in basal diameter and with a thickness of over 3 mm can be sampled transsclerally, mainly if anteriorly located [20] (fig. 1–3). Meticulous

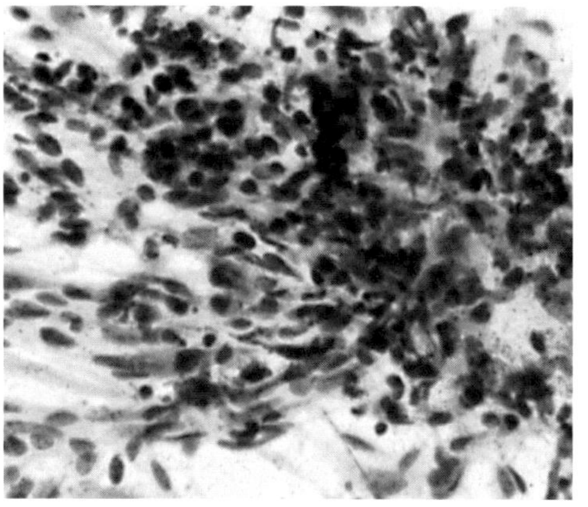

Fig. 1. FNAB of a pigmented choroidal lesion. Cytological examination revealed spindle cells with prominent nucleoli, consistent with spindle-cell choroidal malignant melanomas (Papanicolau ×125).

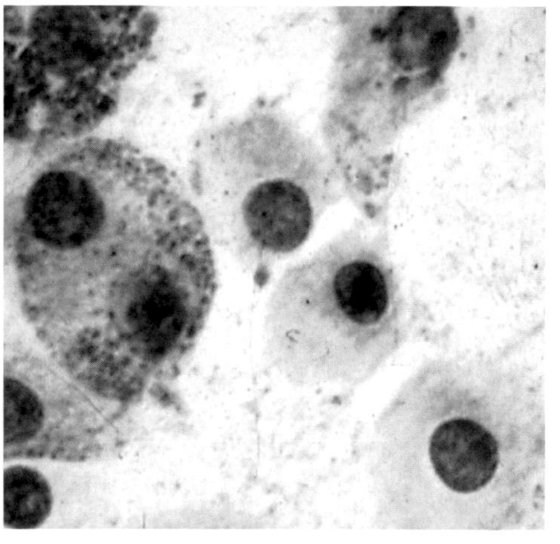

Fig. 2. FNAB of a pigmented ciliary body lesion. Cytological examination revealed round and monomorphic neoplastic cells with prominent nucleoli and abundant melanin, consistent with epithelioid choroidal malignant melanomas (Papanicolau ×630).

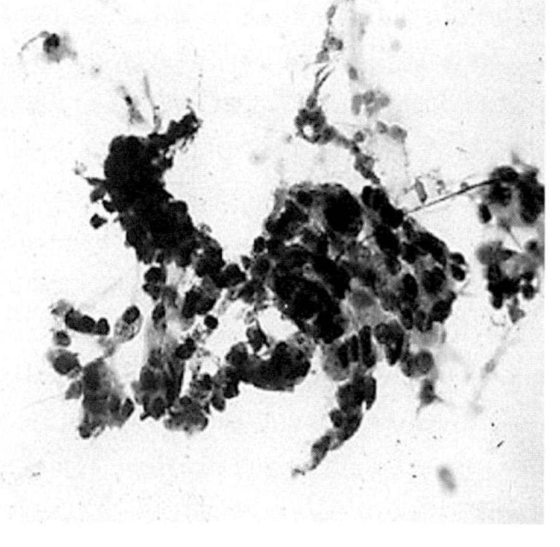

Fig. 3. FNAB of an amelanotic choroidal tumor in an apparently healthy 45-year-old female. Cytological examination revealed cohesive, monomorphic epithelial neoplastic cells with prominent nucleoli, consistent with breast metastatic tumor to the choroid (Papanicolau ×125).

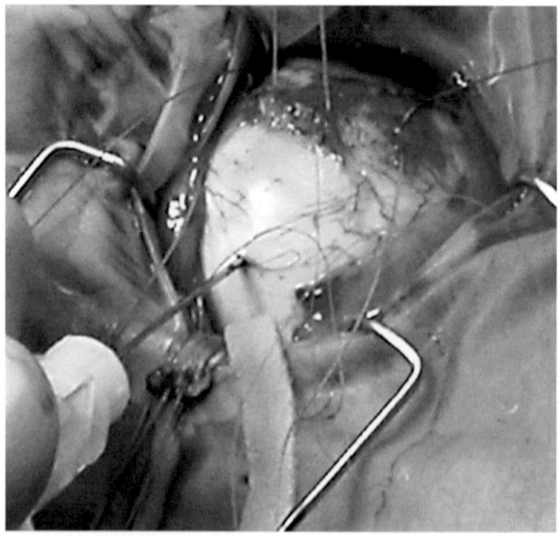

Fig. 4. Transscleral FNAB: note that in this case a transscleral approach is safely applied to a relatively posteriorly located tumor.

localization by both transillumination and indirect ophthalmoscopy is mandatory. Shields et al. [21] obtained adequate material for cytological diagnosis in 88% of patients affected by different intraocular lesions. Uveal melanoma was diagnosed in 38%, metastasis in 20%, retinoblastoma and leukemia in 2%, and lymphoma in 3%. Small lesions with posterior location are usually better biopsied through a transvitreal approach [8]. There is a strong correlation between the results of FNAB and the thickness of the lesions on A-scan ultrasound (<1.9 mm: 40%; 1.9–4 mm: 90%; >4 mm: 98%) [22]. Augsburger et al. [4] investigated FNAB for the cytological diagnosis of small indeterminate choroidal melanocytic lesions, obtaining sufficient material for cytodiagnosis in 65% of tumors whose thickness varied between 1.5 and 3 mm. The most important limitation of FNAB in small tumors is its frequent inability to obtain sufficient cells for diagnosis [2, 5, 23].

The basic equipment required for FNAB is simple: a fine needle (25–30 G), a 10-ml disposable syringe, some glass slides with a frosted-end and a fixative [1]. Different gauges are currently used when performing intraocular tumor biopsy in a routine clinical setting. Most authors prefer a 25-gauge needle for both transvitreal and transscleral approaches [5, 12, 24, 25]. Other authors recommend a 30-gauge needle in transscleral procedures and a 27-gauge instrument for the transvitreal approach [6, 8, 26, 27]. Our standard transscleral FNAB approach encompasses a 25-gauge (25 mm in length) spinal needle connected by a hollow tube to a 10-ml syringe [5, 23]. The needle is inserted into the tumor through a 300-μm scleral incision (to avoid excessive pressure when penetrating the eye). This specific procedure allows safe sampling of small lesions (thickness <3.5 mm), avoiding retinal damage or intravitreous penetration by the needle. A double- or triple-pass sampling with the specific needling procedure is mandatory in all cases [1, 5] (fig. 4). The scleral incision is immediately sutured after sampling and, when performed for cytogenetic

Table 1. Safety and adequacy of FNAB for cytogenetic testing in uveal melanoma

Reference	Eyes	Sampling	Follow-up months	Adequacy	Safety (local)	Safety (general)
Midena et al. [5], 2008	32	25 G transscleral	47	87%	0%	100%
Shields et al. [6], 2007	140	30 G transscleral 27 G transvitreal	8	86%	46% of transient vitreous hemorrhages	100%
Young et al. [27], 2007	18	30 G transsclerral	–	50%	5% of transient hemorrhages	–
Shields et al. [8], 2007	46 tumors measuring 3 mm or less in thickness	30 G transscleral 27 G transvitreal	10	94%	55% of transient vitreous hemorrhages	100%
Shields et al. [51], 2011	500	27 G transvitreal or transscleral	36	92%	drop of preretinal blood at the site of retinal perforation (common but not quantified)	–

testing at the time of brachytherapy, a radioactive plaque is promptly placed over the tumor base not only to treat the tumor, but also to sterilize the needle tract [5, 23].

For transvitreal access, the needle should be bent 2–3 mm from its bevelled tip to an angle of 60–90° relative to the shaft [20]. This allows entry into the neoplasm with the needle tip parallel to the sclera through one of the sclerotomies, reducing the risk for posterior scleral perforation in small tumors. Some authors prefer indirect ophthalmoscopic guidance [6], whereas others use an operating microscope combined with an irrigating contact lens, but without an intraocular fiber-optic light source [22]. Biomicroscopy and wide-angle viewing allow the needle to be placed easily and safely [28]. A transvitreal approach has recently been investigated for posteriorly located tumors in a series of 140 eyes [6]. According to Shields et al. [6], tumors posterior to the equator (67 of 140 eyes: 48%) were sampled via the pars plana, yielding sufficient material in 65 of 67 cases (97%). Unfortunately, a high number of local complications were reported (vitreous hemorrhage in 64 eyes), mainly related to the direct transretinal approach to the tumor apex. In the same series, a lower rate of sufficient material (75 vs. 97%) was obtained when the tumors were sampled with a 30- versus a 27-gauge needle. Balancing needle diameter versus tumor approach needs further investigation (table 1).

Vitrectomy-Based Biopsies

The vitrectomy-based biopsy proposed by Bechrakis et al. [17] encompasses standard 3-port pars plana vitrectomy followed by core vitrectomy (21-gauge) to allow direct access to the tumor. Subsequently, a vitreous separation is induced over the tumor and a thorough vitrectomy is performed over the intended biopsy site (this is believed to avoid vitreoretinal incarceration with the vitreous cutter during the biopsy procedure). The intraocular pressure is then elevated to 70 mm Hg, and the retina overlying the tumor incised by a sharp intraocular Sato knife just to allow the transretinal entrance of the 0.9-mm-thick vitreous cutter. The vitreous cutter is then inserted into the tumor through the retinotomy incision with a high aspiration setting (400 mm Hg) and a low cutting frequency (80/min). No cryotherapy or laser treatment is usually performed at the retinotomy site, and after fluid/gas exchange, 20% SF6 gas is usually injected into the eye and the scleral incisions are closed [17]. In this series (23 tumors), the main reported complication was vitreous hemorrhage in 2 cases (6%) [17]. An inconclusive biopsy result was also reported in 1 case (3%) [17]. In addition 1 multifocal, intraocular tumor spread was documented. No orbital recurrence or metastases were reported 42 months after enucleation. Eight of 34 patients were enucleated because of choroidal melanomas. The authors did not report the reason for enucleation, raising questions about the real safety of this procedure [17].

Vitrectomy using the 25-gauge system was first described in 2002, encouraging some authors to use this technology for choroidal tumor biopsy [29]. In 2006, Sen et al. [30] reported 14 patients undergoing choroidal tumor biopsy using the 25-gauge system approach. The infusion port, light pipe, and vitreous cutter were inserted 4 mm from the limbus and the vitreous cutter was advanced across the vitreous cavity, through the retina, into the tumor. Tissue samples were taken by rotating the cutter within the tumor. Then, the vitreous cutter was repeatedly withdrawn from the tumor and flushed with a small volume of vitreous to prevent blockage of the aspiration cannula by tumor fragments. Total vitrectomy was not performed. When the scleral ports were removed, cotton buds were used to apply pressure to each entry site. Gas tamponade was not used. The authors reported a positive tissue diagnosis in 13 of 14 patients [30]. The conclusion was that performing a choroidal tumor biopsy with a 25-gauge vitreous cutter is associated with a low ocular morbidity and, in most cases, a larger tumor yield compared to FNAB. However, the authors did not report tumor locations and dimensions, making impossible any comparison with standard FNAB [30].

The rationale for performing intraocular biopsy by vitrectomy is that tumor samples obtained by FNAB may be very small, requiring access to a specialized cytology laboratory with an experienced cytopathologist. Unfortunately, compared to the FNAB procedure, this technique is more aggressive, more surgically demanding and its safety remains unknown. Moreover, in our experience, most tumors sampled by vitrectomy procedures could theoretically be managed by standard FNAB [5, 23, 31].

Pars Plana Vitrectomy-Assisted Incisional Biopsies
Kvanta et al. [31] reported on a series of 10 cases that were biopsied with a standard intraocular forceps after performing a diamond knife incision through the pars plana and a vitrectomy. Histological diagnosis was obtained in all cases, leading to subsequent enucleation in 5 tumors (thickness 3.1–6.9 mm). The reasons for enucleation in these medium-sized melanomas, which are commonly amenable to conservative treatment, were not given. A high frequency of postoperative complications was noted. In the 5 nonenucleated eyes, 1 rhegmatogenous retinal detachment, 1 increasing serous retinal detachment and 1 vitreous hemorrhage, which resolved spontaneously before enucleation, were reported [31]. Moreover, some pathological diagnoses in this paper were questioned in a subsequent editorial note (2 cases classified as 'posterior scleritis' on choroidal biopsy and 2 cases classified as 'hematoma') [9].

Safety

Intraocular tumor biopsy can cause different complications [11]. The potential risks, including hemorrhage, retinal detachment, cataract and endophthalmitis, as well as tumor seeding, tumor recurrence or extraocular spread, are well known. However, reliable data to define their incidence and association with different sampling procedures are currently unavailable [11]. Moreover, complications are typically underreported in the literature, reducing the possibility to correctly evaluate the risks and perform a correct comparison between different biopsy methods [11].

It is well known that sampling procedures in retinoblastoma, a poorly cohesive tumor, have a high risk of cell spreading. Notwithstanding, there are no reports of extraocular growth of retinoblastoma using FNAB with 25-gauge or smaller needles [12, 15, 24, 32]. On the other hand, vitrectomy procedures to identify the nature of a retinal mass in childhood are highly dangerous, as reported by Decaussin et al. [33]. The consequences of vitrectomy in cases of unsuspected intraocular tumors in childhood are often disastrous [34]. These data should be taken into account when planning different sampling techniques for any intraocular tumor, mainly when considering the extensive application in a routine clinical setting.

Glasgow et al. [35] first documented sheets and numerous single malignant uveal melanoma cells in the needle tract of enucleated eyes after a transscleral approach using a 30-gauge needle. However, uveal melanoma implantation with subsequent metastasis has never been reported with needles of 25 G or less [36]. Moreover, the risk of needle tract seeding may be reduced to virtually zero by immediate irradiation of the sampling area [23]. In 1996, Char et al. [37] found no difference in survival comparing 116 irradiated melanoma patients previously sampled by FNAB versus 731 unbiopsied patients. Nevertheless, opening of the sclera with a trephine or using a large needle resulted in tumor extension in 6 of 17 cases [38]. Orbital recurrences and death from metastases occurred in 10 of 17 cases within 8 years, although these eyes were enucleated immediately or within 3 days after the biopsy [38].

Transvitreal FNAB is claimed to have a lower risk of tumor spread compared with the transscleral approach. The number of implanted tumor cells appears to be lower than that required to establish tumor implantation under experimental conditions [39]. In addition, according to experimental animal studies, survival of tumor cells in the vitreous is difficult [40]. Nevertheless, 1 suspected (not adequately proven) clinical case of extrascleral recurrence has recently been reported after diagnostic FNAB in uveal melanoma [3]. In this case, a transvitreal approach was used, but no treatment was applied over the sampled area. In contrast, to the best of our knowledge, no extrascleral extension after transscleral FNAB has been documented yet [3]. Moreover, a transscleral approach through the tumor base can be easily sterilized by standard plaque brachytherapy, whereas the transvitreal approach may cause seeding in untreated scleral areas [23]. Given the frequent application of FNAB in virtually every organ of the body, the risk of tumor cell dissemination by fine needle biopsies is extremely low [6]. Moreover, among all the biologic steps related to the establishment of metastatic disease, it is unlikely that mini-invasive tumor sampling techniques significantly affect patient outcome [6, 8]. In contrast, there are only few reports on vitrectomy-based procedures for taking a uveal melanoma biopsy. Therefore, there is too little information to make a valid comparison regarding the relative merits, safety and adequacy of these procedures [11]. Furthermore, other invasive surgical procedures for intraocular tumor sampling (ab externo and ab interno tumor resection) must not be included in the standard armamentarium of uveal melanoma biopsy techniques, but should eventually be considered as alternative treatment options for selected, large uveal melanomas [41, 42]. Safety and efficacy of these procedures remain unknown [42].

Biopsy for Cytogenetic Analysis

Critically looking at uveal melanoma management and treatment results, some unresolved issues still exist. First, we are still unable to accurately determine clinically which lesions need to be treated and we continue to treat each patient affected by small but malignant 'indeterminate choroidal melanocytic lesions' too late [4]. Second, we are still unable to clinically accurately select which patients need systemic adjuvant therapy at the time of diagnosis, that, theoretically, may be more effective in treating microscopic rather than macroscopic tumor metastases, where multiple mechanisms of resistance can develop [23]. Third, clinically oriented cytogenetic analysis is currently considered the most sensible and specific tool to obtain patient prognostication in uveal melanoma, but it is not routinely used in most centers (fig. 4, 5) [23]. All these considerations suggest a detailed reevaluation of the role of biopsies in uveal melanoma management. In our opinion, there are two main reasons to avoid taking a tumor biopsy for cytogenetic analysis in a routine clinical setting: (1)

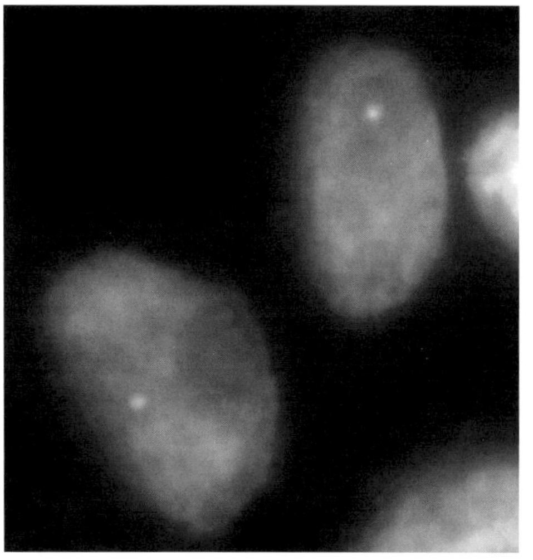

Fig. 5. FNAB in a case of posterior choroidal melanoma for cytogenetic testing. FISH revealed monosomy 3 of the lesion (note the single white dot inside each cell). This patient developed liver metastatic disease 3 years after ^{125}I brachytherapy.

there are currently no proven adjuvant systemic therapies for high-risk patients (this concept violates the general indication for intraocular tumor biopsy) [5, 23] and (2) major complications can ensue [3]. Conversely, there is a single, but strong indication to routinely introduce this procedure: that the use of new drugs as adjuvant therapy in nonselected populations often has a limited effectiveness [43]. In a scenario where new drugs for systemic treatment develop continuously, it is mandatory to identify subgroups of patients amenable to receive a 'tailored' optimal treatment [5, 23, 43].

In addition to these general considerations, specific issues exist for cytogenetic sampling of small indeterminate melanocytic proliferations [8]. First, complications are more common in small lesions compared to large ones. Second, we know cytogenetic differences between low-risk versus high-risk tumors, but we do not know cytogenetic differences between nevi and premalignant lesions (lesions that need to be treated before becoming biologically malignant) [5]. Third, we do not know if treatment of any small but malignant uveal melanocytic proliferation prolongs survival [4]. According to cancer biology, much of the improvement in survival rates associated with early detection is due to the treatment of a higher proportion of premetastatic lesions, not from curing malignancies [4]. Evidence supports the concept that it is likely that uveal melanoma forms micrometastases in susceptible patients 5 years prior to ocular tumor diagnosis (and biopsy) [44].

Heterogeneity

Tumor heterogeneity is considered a consequence of cancer pathogenesis [45]. Cancer development is often associated with genomic instability and acquisition of genomic heterogeneity generating both clonal and nonclonal tumor cell populations [45]. Morphologic heterogeneity is well recognized in uveal melanoma showing variable proportions of epithelioid and spindle cells [46]. Epithelioid cells more often show loss of chromosome 3 than spindle cells [46]. Sandinha et al. [47] and White et al. [48] reported cytogenetic heterogeneity in uveal melanoma, describing morphologic heterogeneity corresponding with cytogenetic heterogeneity. In the case of FNAB sampling, intratumor heterogeneity may interfere with a correct prediction of the patient's prognosis [46–48].

In our clinical series, some eyes without loss of chromosome 3 developed metastatic disease. It is possible that these tumors evolved in a different manner, but it may also be due to our inability to detect, for example, partial loss of chromosome 3, isodisomy 3 (duplication of one copy of a chromosome), or it may be due to intratumor heterogeneity [43]. Maat et al. [49] pointed out that heterogeneity in monosomy 3 is a frequent event in uveal melanoma. In their study, 7 of 50 tumors were found to be heterogeneous, but the FISH technique was unsuccessful in 17 lesions and difficult to be interpreted in 8. Furthermore, the FISH technique was performed for chromosome 3 only, without a control probe for aneuploidy and/or truncation artifacts. Dopierala et al. [50] have also reported that approximately 50% of the examined tumors showed heterogeneity of at least one locus on chromosome 3. Such a level of heterogeneity may affect cytogenetic prognostication in a large portion of uveal melanoma patients, mainly when biopsied by FNAB. In contrast, these data do not match with the predictive value of monosomy 3 obtained in vivo by FNAB specimens, nor with clinically demonstrated different life expectancy of uveal melanoma patients with and without this cytogenetic alteration [5, 6, 8]. Mensink et al. [46] have recently investigated the presence of focal or diffuse chromosome 3 intratumoral heterogeneity in uveal melanoma using direct interphase FISH in a series of 151 uveal melanomas (the largest series ever reported). Tumors with monosomy 3 were suspected to be heterogeneous if there were low percentages of monosomy 3, triploid clones, inconsistencies between FISH on centromere 3 and the long arm of chromosome 3, or discrepancies between FNAB and the main tumor. Only such tumors (n = 16) were selected and analyzed for intratumor heterogeneity. Six tumors showed the same percentage of monosomy 3 throughout the tumor, and 10 showed multiple clones with different percentages of monosomy 3. However, these tumors did not show focal heterogeneity with respect to chromosome 3 status (i.e. monosomy vs. disomy), and differences in monosomy 3 distribution between the base and apex of the tumor could not be identified. We agree with Mensink et al. [46] concluding that although a small number of uveal melanomas may show heterogeneity for chromosome 3, it does not affect the patient's prognosis. This evidence is also supported by extensive clinical application results [5, 6, 8]. In the presence of triploid clones or uncommon abnormalities, chromosome 3 loss is more difficult to be interpreted, but most uveal melanoma biopsies provide an accurate prediction of patient's prognosis, mainly when associated with classical clinical prognostic factors [46, 51]. Nevertheless, a multiple sampling procedure (through the same access), with separate specimen collection and analysis, is always recommended [5]. Moreover, multichromosomal analysis techniques, such as multiplex ligation-dependent probe amplification or multichromosomal FISH, better characterize patients' prognosis [43, 52].

Conclusion

All currently available methods for intraocular tumor biopsy differ by: amount of specimen, frequency of an insufficient specimen, potential for a sampling error, morbidity related to the procedure, potential for tumor seeding, costs and the skill and expertise required from both the surgeon performing the biopsy and the pathologist analyzing and interpreting the obtained specimens [9]. There is still limited information to make a reliable comparison among all biopsy procedures [9, 11]. However, less invasive methods with (1) high safety, (2) acceptable adequacy and (3) low costs are the best candidates for extensive clinical application [5, 23]. Ophthalmic oncology is now rapidly moving in the same direction as all other oncologic subspecialties, starting to determine the patient's risk using sampling-based laboratory techniques. However, it is essential to start treating patients more often in the premalignant tumor stage and to develop consistent, selective and effective adjuvant therapy [4, 23].

References

1 Midena E, Segato T, Piermarocchi S, Boccato P: Fine needle aspiration biopsy in ophthalmology. Surv Ophthalmol 1985;29:410–422.

2 The Collaborative Ocular Melanoma Study Group: Accuracy of diagnosis of choroidal melanoma in the Collaborative Ocular Melanoma Study. COMS report No 1. Arch Ophthalmol 1990;108:1268–1273.

3 Caminal JM, Sanz S, Carreras M, Català I, Arruga J, Rocca G: Epibulbar seeding at the site of a transvitreal fine-needle aspiration biopsy. Arch Ophthalmol 2006;124:587–589.

4 Augsburger JJ, Correa ZM, Schneider S, Yassin RS, Robinson-Smith T, Ehya H, Trichopoulos N: Diagnostic transvitreal fine-needle aspiration biopsy of small melanocytic choroidal tumours in nevus versus melanoma category. Trans Am Ophthalmol Soc 2002;100:225–232.

5 Midena E, Bonaldi L, Parrozzani R, Radin PP, Boccassini B, Vujosevics S: In vivo monosomy 3 detection of posterior uveal melanoma: 3-year follow-up. Graefes Arch Clin Exp Ophthalmol 2008;246:609–614.

6 Shields CL, Ganguly A, Materin MA, Teixeira L, Mashayekhi A, Swanson LA, Marr BP, Shields JA: Chromosome 3 analysis of uveal melanoma using fine-needle aspiration biopsy at the time of plaque radiotherapy in 140 consecutive cases: the Deborah Iverson, MD, Lectureship. Arch Ophthalmol 2007;125:1017–1024.

7 The Collaborative Ocular Melanoma Study Group: The COMS randomized trial of iodine 125 brachytherapy for choroidal melanoma. 3. Initial mortality findings. COMS Report No 18. Arch Ophthalmol 2001;119:969–982.

8 Shields CL, Materin MA, Teixeira L, Mashayekhi A, Ganguly A, Shields JA: Small choroidal melanoma with chromosome 3 monosomy on fine-needle aspiration biopsy. Ophthalmology 2007;114:1919–1924.

9 Augsburger JJ: Diagnostic biopsy of selected intraocular tumors. Am J Ophthalmol 2005;140:1094–1095.

10 Faulkner-Jones BE, Foster WJ, Harbour JW, Smith ME, Davila RM: Fine-needle aspiration biopsy with adjunct immunohistochemistry in intraocular tumour management. Acta Cytol 2005;49:297–308.

11 Eide N, Walaas L: Fine-needle aspiration biopsy and other biopsies in suspected intraocular malignant disease: a review. Acta Ophthalmol 2009;83:588–601.
12 Eide N, Syrdalen P, Walaas L, Hagmar B: Fine-needle aspiration biopsy in selecting treatment for inconclusive intraocular disease. Acta Ophthalmol Scand 1999;77:448–452.
13 All-Ericsson C, Economou MA, Landau I, Traisk F, Seregard S: Uveitis masquerade syndromes: diffuse retinoblastoma in an older child. Acta Ophthalmol Scand 2007;85:569–570.
14 Woog JJ, Chess J, Albert DM, Dueker DK, Berson FG, Craft J: Metastatic carcinoma of the iris simulating iridocyclitis. Br J Ophthalmol 1984;68:167–173.
15 Char DH, Kemlitz AE, Miller T, Crawford JB: Iris ring melanoma: fine-needle biopsy. Br J Ophthalmol 2006;90:420–422.
16 Shields CL, Manquez ME, Ehya H, Mashayekhi A, Danzig CJ, Shields JA: Fine-needle aspiration biopsy of iris tumours in 100 consecutive cases: technique and complications. Ophthalmology 2006;113:2080–2086.
17 Bechrakis NE, Foerster MH, Bornfeld N: Biopsy in indeterminate intraocular tumours. Ophthalmology 2002;109:235–242.
18 Pe'er J, Blumenthal EZ, Frenkel S: Punch biopsy of iris lesions: a novel technique for obtaining histology samples. Br J Ophthalmol 2007;91:660–662.
19 Finger PT, Latkany P, Kurli M, Iacob C: The Finger iridectomy technique: small incision biopsy of anterior segment tumours. Br J Ophthalmol 2005;89: 946–949.
20 Eide N, Syrdalen P, Walaas L, Hagmar B: Fine-needle aspiration biopsy in selecting treatment for inconclusive intraocular disease. Acta Ophthalmol Scand 1999;77:448–452.
21 Shields JA, Shields CL, Ehya H, Eagle RC Jr, De Potter P: Fine-needle aspiration biopsy of suspected intraocular tumours. The 1992 Urwick Lecture. Ophthalmology 1993;100:1677–1684.
22 Cohen VM, Dinakaran S, Parsons MA, Rennie IG: Transvitreal fine-needle aspiration biopsy: the influence of intraocular lesion size on diagnostic biopsy result. Eye 2001;15:143–147.
23 Midena E, Bonaldi L, Parrozzani R, Tebaldi E, Boccassini B, Vujosevic S: In vivo detection of monosomy 3 in eyes with medium-sized uveal melanoma using transscleral fine needle aspiration biopsy. Eur J Ophthalmol 2006;16:422–425.
24 Augsburger JJ, Shields JA, Folberg R, Lang W, O'Hara BJ, Claricci JD: Fine-needle aspiration biopsy in the diagnosis of intraocular cancer. Cytologic-histologic correlations. Ophthalmology 1985;92:39–49.
25 Char DH, Miller TR, Ljung BM, Howes EL Jr, Stoloff A: Fine-needle aspiration biopsy in uveal melanoma. Acta Cytol 1989;33:599–605.
26 Young TA, Burgess BL, Rao NP, Gorin MB, Straatsma BR: High-density genome array is superior to fluorescence in situ hybridization analysis of monosomy 3 in choroidal melanoma fine-needle aspiration biopsy. Mol Vis 2007;13:2328–2333.
27 Young TA, Burgess BL, Rao NP, Glasgow BJ, Straatsma BR: Transscleral fine-needle aspiration biopsy of macular choroidal melanoma. Am J Ophthalmol 2008;145:297–302.
28 Robertson DM: Cytogenetics in the management of uveal melanoma: are we there yet? Arch Ophthalmol 2008;126:409–410.
29 Fujii GY, De Juan E Jr, Humayun MS: Initial experience using the transconjunctival sutureless vitrectomy system for vitreoretinal surgery, Ophthalmology 2002;109:1814–1820.
30 Sen J, Groenewald C, Hiscott PS, Smith PA, Damato BE: Transretinal choroidal tumor biopsy with a 25-gauge vitrector. Ophthalmology 2006;113:1028–1031.
31 Kvanta A, Seregard S, Kopp ED, All-Ericsson C, Landau I, Berglin L: Choroidal biopsies for intraocular tumours of indeterminate origin. Am J Ophthalmol 2005;140:1002–1006.
32 O'Hara BJ, Ehya H, Shields JA, Augsburger JJ, Shields CL, Eagle RC Jr: Fine-needle aspiration biopsy in paediatric ophthalmic tumours and pseudotumours. Acta Cytol 1993;37:125–130.
33 Decaussin M, Boran MD, Salle M, Grange JD, Patricot LM, Thivolet-Bejui F: Cytological aspiration of intraocular retinoblastoma in an 11-year-old boy. Diagn Cytopathol 1998;19:190–193.
34 Stevenson KE, Hungerford J, Garner A: Local extraocular extension of retinoblastoma following intraocular surgery. Br J Ophthalmol 1989;73:739–742.
35 Glasgow BJ, Brown HH, Zargoza AM, Foos RY: Quantitation of tumour seeding from fine-needle aspiration of ocular melanomas. Am J Ophthalmol 1988;105:538–546.
36 Char DH, Miller T: Accuracy of presumed uveal melanoma diagnosis before alternative therapy. Br J Ophthalmol 1995;79:692–696.
37 Char DH, Kroll SM, Miller T, Castro J, Quivey J: Irradiated uveal melanomas: cytopathologic correlation with prognosis. Am J Ophthalmol 1996;122: 509–513.

38 Jensen OA, Andersen SR: Late complications of biopsy in intraocular tumours. Acta Ophthalmol 1959;37:568–575.
39 Eriksson O, Hagmar B, Ryd W: Effects of fine-needle aspiration and other biopsy procedures on tumour dissemination in mice. Cancer 1994;54:73–78.
40 Snell AC Jr: Heterotransplantation of tumours into various regions of the guinea-pig eye. Am J Ophthalmol 1951;34:733–738.
41 Gunduz K, Bechrakis NE: Exoresection and endoresection for uveal melanoma. Middle East Afr J Ophthalmol 2010;17:210–216.
42 Singh AD, Triozzi PL: Endoresection for choroidal melanoma: palliative or curative intent? Br J Ophthalmol 2008;92:1015–1016.
43 Bonaldi L, Midena E, Filippi B, Tebaldi E, Marcato R, Parrozzani R, Amadori A: FISH analysis of chromosomes 3 and 6 on fine needle aspiration biopsy samples identifies distinct subgroups of uveal melanomas. J Cancer Res Clin Oncol 2008;134:1123–1127.
44 Eskelin S, Pyrhönen S, Summanen P, Hahka-Kemppinen M, Kivelä T: Tumor doubling times in metastatic malignant melanoma of the uvea: tumor progression before and after treatment. Ophthalmology 2000;107:1443–1449.
45 Bayani J, Selvarajah S, Maire G, et al: Genomic mechanisms and measurement of structural and numerical instability in cancer cells. Semin Cancer Biol 2007;17:5–18.
46 Mensink HW, Vaarwater J, Kilic E, et al: Chromosome 3 intratumor heterogeneity in uveal melanoma. Invest Ophthalmol Vis Sci 2009;50:500–504.
47 Sandinha T, Farquharson M, McKay I, Roberts F: Correlation of heterogeneity for chromosome 3 copy number with cell type in choroidal melanoma of mixed-cell type. Invest Ophthalmol Vis Sci 2006;47:5177–5180.
48 White VA, McNeil BK, Thiberville L, Horsman DE: Acquired homozygosity (isodisomy) of chromosome 3 during clonal evolution of a uveal melanoma: association with morphologic heterogeneity. Genes Chromosomes Cancer 1996;15:138–143.
49 Maat W, Jordanova ES, van Zelderen-Bhola SL, et al: The heterogeneous distribution of monosomy 3 in uveal melanomas: implications for prognostication based on fine-needle aspiration biopsies. Arch Pathol Lab Med 2007;131:91–96.
50 Dopierala J, Damato BE, Lake SL, Taktak AF, Coupland SE: Genetic heterogeneity in uveal melanoma assessed by multiplex ligation-dependent probe amplification. Invest Ophthalmol Vis Sci 2010;51:4898–4905.
51 Shields CL, Ganguly A, Bianciotto CG, Turaka K, Tavallali A, Shields JA: Prognosis of uveal melanoma in 500 cases using genetic testing of fine-needle aspiration biopsy specimens. Ophthalmology 2011;118:396–401.
52 Damato B, Dopierala JA, Coupland SE: Genotypic profiling of 452 choroidal melanomas with multiplex ligation-dependent probe amplification. Clin Cancer Res 2010;16:6083–6092.

Edoardo Midena, MD, PhD
Department of Ophthalmology, University of Padova
Via Giustiniani 2
IT–35128 Padova (Italy)
Tel. +39 049 821 2121, E-Mail edoardo.midena@unipd.it

Jager MJ, Desjardins L, Kivelä T, Damato BE (eds): Current Concepts in Uveal Melanoma.
Dev Ophthalmol. Basel, Karger, 2012, vol 49, pp 96–116

Analysis of Intraocular Biopsies

Sarah E. Coupland

Department of Cellular and Molecular Pathology, University of Liverpool, Liverpool, UK

Abstract

In this chapter, the importance of intraocular biopsies in the diagnosis/exclusion of ocular malignancies and prognostication is outlined. Despite improvements in ancillary studies in ophthalmology, intraocular biopsies are increasingly being performed in many ocular oncology centres. Experience is required in taking these biopsies, in their transport to the pathology laboratory, in their triaging and processing, and in their interpretation. To optimize the biopsy yield, well-tested and practical standard operational procedures for morphological, immunocytological and molecular genetic analyses are necessary. 'Tips and tricks' for the fixation and processing of intraocular tissue and fluid biopsies are provided. For example, a fixative such as Cytolyt or HOPE fixation is recommended for vitreous biopsies, allowing for all investigations to be performed, including DNA-based molecular genetic techniques, such as polymerase chain reaction used in clonality analysis in lymphoma diagnosis and in organism identification in endophthalmitis. Most solid tissue samples can be placed in buffered formalin, with the exception of those requiring RNA-based molecular techniques, here fresh tissue being preferable. The importance of incorporating data from all investigations and summarizing them as an integrated report is emphasized. The pitfalls of using any single test (e.g. a molecular genetic test) as a 'stand-alone' investigation are highlighted. Communication with relevant clinical information between surgeon and pathologist is essential at each stage: for sample delivery, for exclusion of differential diagnoses and for rapid result transmission. The current molecular genetic techniques for uveal melanoma prognostication are summarized, and how their data can be used for instigation of individualized management plans for patients discussed.

The treatment of malignant disease usually requires a diagnosis based on microscopical examination of tumour tissue. In ocular oncology, however, the diagnosis of an intraocular malignancy is often based on clinical examination together with comprehensive non-invasive ancillary tests, such as ultrasonography or angiography, and occasionally serological tests. In some cases, however, lesions which appeared as 'typical' for a particular ocular malignancy have been diagnosed as something quite different following cytopathological diagnosis [1, 2]. Intraocular biopsies are increasingly being performed in ocular oncology, not only for diagnosis confirmation but also for prognostication. In this way, individualized management plans for each patient can be instigated. Various surgical methods have been applied since Hirschberg first performed an intraocular biopsy in 1868 [3]. These surgical procedures will not be the subject of this chapter, and the readers are referred to relevant reviews [4–11]. Essentially, the bioptic targets include all intraocular structures, such as the iris, ciliary body, choroid, retina as well as the aqueous and vitreous humour. Samples from these tissues can be obtained via translimbal, transscleral, transvitreal and/or transretinal approaches. Biopsy procedures are aspirational (using a fine needle or a vitreous cutter), incisional or excisional (e.g. local tumour resection or enucleation, if required) [10]. All these procedures are potentially hazardous and have certain failure rates; however, the various risks with which they are associated differ and are also dependent on the experience of both the surgeon and reporting pathologist. The aims of this chapter are to: (1) provide an overview of intraocular biopsies from the pathologist's perspective; (2) discuss specimen preparation and processing to obtain maximum yield; (3) consider the differential diagnoses in the particular ocular structures, and (4) describe the various methodologies (including cytogenetic and molecular genetic techniques) that can be applied.

Sample Handling and Processing

Since intraocular tumours may be located in differing ocular tissues, such as the uvea or the retina, with possible involvement of the anterior chamber or vitreous, the histopathologist is confronted with a variety of specimens, including aqueous taps, vitreous taps, diagnostic vitrectomy specimens, uveal biopsies and subretinal aspirates. Depending on the clinical question, the biopsy type and the time required for transport to the diagnostic laboratory, intraocular tissue biopsies should be sent to the laboratory either fresh or in a fixative. It is, therefore, advisable that the surgeon discusses the case with the investigating laboratory before the specimen is collected, so that the correct fixative and container are used and to make any special arrangements for transport. On receipt of the specimen in the diagnostic laboratory, careful consideration should be given to the handling and examination of the samples, incorporating a triage system, allowing for the application of cytological, immunocytological, and molecular biological analyses, in order to optimize the diagnostic and prognostic yields [12–14].

Intraocular Fluid Biopsies and Aspirational Tumour Biopsies

In the case of intraocular fluid biopsies and tumour fine needle aspiration biopsies (FNAB), it is preferable that these are delivered to the investigating laboratory without fixative, if possible within 1 hour of the procedure. However, if longer delays are anticipated, for example, if the sample is being assessed at a remote laboratory, the specimen should be placed in culture medium (e.g. bovine serum albumin) or in a mild cytofixative, such as herpes/glutamic acid buffer-mediated organic solvent protection effect (HOPE) fixation or Cytolyt (Cytyc) for subsequent ThinPrep slide preparation [15, 16]. The latter two fixatives are preferable to formalin, glutaraldehyde or alcohol fixation because they provide superior preservation of cytomorphology and immunoreactivity, as well as for possible DNA extraction for some of the molecular genetic methods described below. Various techniques have been described in the literature for the preparation of intraocular fluid (vitreous and aqueous) specimens for cytomorphological evaluation [17–22]. These include: (1) vitreous 'filtration'; (2) a celloidin bag technique; (3) cytospin, and (4) cell block preparation. We are most familiar with the latter two methods, whereby the vitreous is spun at 500 rpm for 5 min, concentrating the cells either onto glass slides or into paraffin [12]. The preparation of tumour FNAB is similar to that of ocular fluids, and depends on the degree of cellularity, which is initially evaluated macroscopically. In the case of paucicellular to moderately cellular specimens, cytospins are prepared using the Shandon or ThinPrep techniques, with one cytospin being stained with May Grunewald Giemsa (MGG) for morphological assessment, and 2–3 unstained cytospins being prepared for appropriate special stains or immunocytology (table 1) [13]. Using the MGG stain, the cytospins are scored for various factors including total cell density, proportion of tumour cells within the sample, degree of pigmentation as well as presence or absence of a haemorrhagic or necrotic background. These scores are particularly important with respect to the concentration and quality of DNA extracted for further molecular genetic analyses [Coupland et al., unpublished].

In biopsies with a large number of cells or with visible cell clumps, a paraffin cell block would be more appropriate to allow for better morphological interpretation following sectioning. Again, depending on the morphological findings observed in the haematoxylin-and-eosin-stained sections of these samples, various additional special stains or immunohistochemical markers may be necessary, as summarized in table 1. In all cases, retention of part of the tumour sample should always be considered for DNA extraction, either from the fresh or fixed sample, in order to allow for future molecular genetic analyses (table 2).

Incisional and Excisional Tumour Biopsies

These larger solid tumour specimens, including incisional biopsy, endoresection, local resection and enucleation, are usually fixed in 10% buffered formalin, and processed in paraffin using standard procedures for sectioning and conventional and immunohistochemical techniques for staining. When required, DNA can be extracted from the formalin-fixed paraffin-embedded material for molecular analyses [12, 23].

Table 1. Useful stains in the assessment of intraocular biopsies

Conventional histochemical stains	Detects	Examples/comments
Morphological stains (e.g. MGG, modified Papanicolaou)		
Giemsa	Bacteria Fungi Parasites Protozoa	*Neisseria* sp. *Actinomyces* sp. Toxoplasmosis *Leishmania tropica*
Gram Twort	Gram-positive bacteria (blue) Gram-negative bacteria (red)	
Periodic acid-Schiff with or without diastase digestion	Bacteria Fungi Parasites	*Tropheryma whipplei* *Candida* sp. *Toxoplasma gondii*
Silver stains (e.g. Gomori's methenamine silver; Grocott)	Fungi	
Mucicarmine	Fungi	Encapsulated fungi e.g. *Cryptococcus neoformans*
(Modified) Ziehl-Neelsen	Mycobacteria Nocardia	*M. tuberculosis* *M. leprae*
Warthin Starry	Bacteria Fungi Spirochaetes	*Helicobacter* sp. *Treponema pallidum*
Perl's Prussian blue	Haemosiderin (ferric iron)	
Congo Red	Amyloid Parasites	Green birefringence *Taenia echinococcus*
Immunohistochemistry		
CD3	Pan T-cell marker	
CD20 and PAX-5	B-cell markers	Negative on precursor B-cells and plasma cells
CD138, MUM1/IRF4	Plasma cell markers	
Kappa and Lamda IgM and IgD		Useful combinations when trying to demonstrate clonality of B-cell and plasma cell infiltrates
CD34, CD117	Haemopoietic blasts	Highlight acute myeloid leukaemic infiltrations

Table 1. Continued

Conventional histochemical stains	Detects	Examples/comments
CD68PG	Pan-macrophage marker	
MelanA, HMB-45, MITF	Melanocyte markers	
Cytokeratin		e.g. MNF-116, AE1/AE3, CAM5.2
CK7 and CK20	Cytokeratin subtypes	Enable closer approximations of likely site of primary carcinoma in intraocular carcinoma metastases
PSA	Prostate specific antigen	Nuclear staining in metastatic prostate carcinoma cells
TTF1	Thyroidal transcription factor 1	Nuclear staining in 80% primary and metastatic pulmonary adenocarcinomas
Anti-cytomegalovirus		Against early and late antigens allowing for detection in nucleus and cytoplasm
Anti-*Tropheryma whipplei*		Polyclonal antibody

Examination of Anterior Segment Samples

Aqueous tap is an option for cellular infiltrates in the anterior chamber. When assessing the cytomorphology of these invariably small samples, the age of the patient should be taken into account, so the appropriate stains can be performed on the limited number of cytospins according to the most likely diagnosis. For example, in younger patients, cellular infiltrates can be seen in juvenile xanthogranulomatosis (fig. 1), undiagnosed retinoblastoma, medulloblastoma (fig. 1) or recurrent or previously undiagnosed acute leukaemia. Consequently, key immunostains in such cases are CD68, vimentin, cytokeratin and myeloperoxidase (and/or CD34), respectively. In an older patient, the neoplastic lesions to be considered in the differential diagnoses include an anterior chamber manifestation of uveal melanoma or vitreoretinal lymphoma, secondary intraocular involvement of systemic lymphoma (e.g. B-cell non-Hodgkin lymphoma or multiple myeloma) or leukaemia, as well as an intraocular metastasis of carcinoma. Consequently, important initial immunostains in these cases to consider would be

Table 2. Methods used in the cytogenetic and molecular genetic analysis of uveal melanoma

Method	Material examined/ required	Summary
Fluorescence in situ hybridization	Tumour chromosome	A cytogenetic technique that is used to detect and localize the presence or absence of specific DNA sequences on chromosomes. FISH uses fluorescent probes that bind to only those parts of the chromosome with which they show a high degree of sequence similarity. For example, centromeric probe on chromosomes 3 and 8, as well as directed against 6p and 8q (c-myc) in uveal melanoma
Multiple ligation dependent probe amplification	Tumour DNA	Multiplex polymerase chain reaction-based method examining numerous particular loci in one reaction. For example, the SALSA kit (P029) examines 31 loci across chr. 1, 3p, 6 and 8 in uveal melanoma
Microsatellite analysis	Tumour DNA and control serum	Microsatellites, also known as simple sequence repeats or short tandem repeats, are repeating sequences of 1–6 base pairs of DNA. They are used as molecular markers to study gene duplication or deletion. Microsatellites can be amplified for identification by polymerase chain reaction, using the unique sequences of flanking regions as primers
Array comparative genomic hybridization	Tumour DNA	Array comparative genomic hybridization is a technique to detect genomic copy number variations at a higher resolution level than chromosome-based comparative genomic hybridization. DNA from a test sample and normal reference sample are labelled differentially, using different fluorophores, and hybridized to several thousand probes. The probes are derived from most of the known genes and non-coding regions of the genome, printed on a glass slide. The ratio of the fluorescence intensity of the test to that of the reference DNA is then calculated, to measure the copy number changes for a particular location in the genome
Gene expression profiling	Tumour RNA	Gene expression profiling is the measurement of the activity (the expression) of thousands of genes at once, to create a global picture of cellular function. By examining the mRNA, the activity of the cell (e.g. tumour cell) can be determined

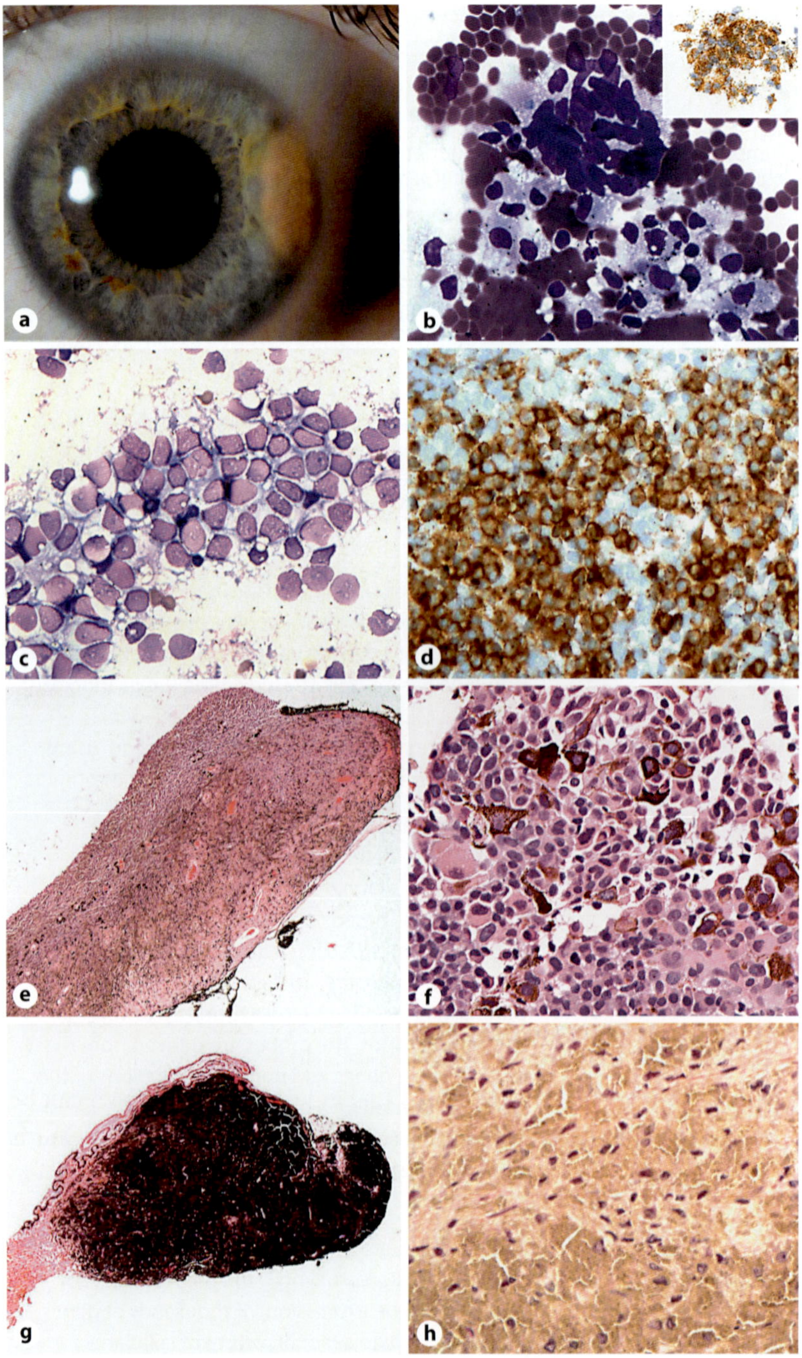

Fig. 1. a Clinical photograph of a yellow-coloured iris lesion in a 16-year-old patient. **b** Cytospins demonstrated that this tumour consisted predominantly of macrophages with foamy cytoplasm on a haemorrhagic background. Although no obvious Touton-like macrophages were present, the findings were consistent with juvenile xanthogranulomatosis (MGG, ×60). Inset: CD68+ macrophages (PAP stain); the MelanA stain was negative. **c** Cytospins of an intraocular biopsy performed in 9-year-old girl with a cystic tumour involving the iris and ciliary body. Numerous immature blasts

MelanA, CD20 (or PAX5), CD138, myeloperoxidase and pancytokeratin (e.g. MNF-116), respectively (table 1). Incisional or excisional biopsies may be performed in patients with iridal lesions of uncertain nature, by using FNAB, vitreous cutter, or by performing either a peripheral iridectomy or iridocyclectomy. Shields et al. [24] presented a review of 100 consecutive cases of iridal lesions biopsied over a 24-year period and reported a diagnostic yield of 99% after performing 1 (74%), 2 (24%) or 3 (2%) passes into the lesions. The advantage of either a peripheral iridectomy or iridocyclectomy over an FNAB or a vitreous cutter is the maintenance of the iris and ciliary body architecture, which is particularly important for the pathologist to relate potentially neoplastic cells to their normal tissue counterparts and to the iris/ciliary body structure. For example, in the differential diagnosis of iris naevus versus spindle cell iris melanoma, it is important to observe the relationship of the proliferating melanocytes with the stroma and with respect to the anterior surface of the iris (fig. 1) [25]. An iridocyclectomy is also useful in cases of melanocytoma (fig. 1) as well as in clarifying puzzling ciliary body or iris tumours, whose clinical response is at variance to that expected. For example, a pigmented ciliary body adenocarcinoma primarily treated as melanoma with proton beam radiotherapy was successfully treated with an iridocyclectomy on development of 'toxic tumour syndrome', revealing the correct diagnosis [26]. Another rare oddity diagnosed with the help of an iridocyclectomy was ectopic thyroid gland in the iris in a 15-year-old male. [27].

Examination of Choroidal Samples

Posterior uveal tumour samples vary considerably in volume between those obtained by FNAB, vitreous cutter, endoresection, local resection and enucleation. The procedure performed depends on a number of factors, including tumour location, tumour size, clinical differential diagnoses and anticipated result, patient age, safety and motivation. There is a strong correlation between the success rate in achieving an unequivocal diagnosis using FNAB and tumour thickness [28, 29]. Essentially, FNAB can be expected to give useful information in choroidal lesions >3 mm in thickness, but is unreliable in lesions thinner than 2 mm. For example, histological differentiation

with narrow cytoplasmic rims and large nuclei with dense chromatin and occasional discrete blasts were seen (MGG, ×40). **d** Immunostains demonstrated positivity for vimentin, cytokeratin and synaptophysin (PAP, ×40). **e** Iridocyclectomy specimen with a densely infiltrating melanocytic tumour with destruction of the architecture of the iris, and with a dense plaque-like growth of tumour cells on the anterior surface of the iris and extending into the angle (not shown) (HE section, ×10). **f** High-power magnification of the tumour cells demonstrated significant cellular atypia, consistent with an iris melanoma (HE section, ×40). **g** Iridocyclectomy specimen of a small dark-pigmented tumour (HE section, ×2). **h** Bleached PAS stain shows that the melanocytic lesion was composed of small naevoid cells with voluminous cytoplasm in the absence of cellular atypia and mitoses, consistent with a magnocellular naevus (melanocytoma) (periodic acid-Schiff stain, ×10).

between uveal naevi and a spindle cell melanoma may be extremely difficult. Our protocol for establishing a diagnosis using morphological and immunohistochemical stains is described above. Essentially, included in the differential diagnosis are primary neoplasms of the choroid (e.g. uveal melanoma, choroidal neurilemmoma, suprachoroidal adenoma, choroidal lymphoma) as well as secondary neoplasms (e.g. metastatic carcinoma, choroidal manifestations of systemic lymphoma and rare metastases of sarcomas) (fig. 2) [30]. When diagnosing uveal melanoma, an attempt at typing with respect to cell morphology is made, although this may be problematic in paucicellular or necrotic specimens. Additionally, when sufficient material is available, immunocytological stains for markers, such as heat shock protein (HSP)-27, are performed. The decreased expression of the latter in uveal melanoma cells with epithelioid cell morphology has been associated with chromosome 3 loss [31]. Whenever possible, DNA is extracted from the biopsy specimen to assess chromosomal status: at present, we are using multiplex ligation-dependent probe amplification (MLPA) to examine 31 loci across chromosomes 1, 3, 6 and 8 [32, 33].

Prognostication and Intraocular Biopsies

Approximately 90% of all uveal melanomas involve the choroid. Despite successful ocular treatment, about 50% of patients die of metastatic disease, which usually involves the liver. Estimation of metastasis and therefore survival probability in uveal melanoma tends to be based on clinical features, particularly largest basal tumour diameter, tumour thickness, ciliary body involvement and extraocular spread. These tumour characteristics form the basis of the 7th edition of the UICC/AJCC TNM staging system. Several histomorphological predictors for metastasis in uveal melanomas are recognized, with the most widely used comprising the presence of epithelioid cells; high mitotic count; low HSP-27 staining score, and the presence of periodic-acid-Schiff-positive closed connective tissue loops [34]. In 1996, Prescher et al. [35] demonstrated a strong correlation between chromosome 3 loss and metastatic death. Since then, other chromosomal abnormalities have been shown to correlate with poor prognosis, including 8q gain, 8p loss, 1p loss, 6q loss and lack of 6p gain [36, 37]. In 1999, we started offering genetic tumour typing to all patients treated by local resection or enucleation using fluorescence in situ hybridization (FISH), examining the status of chromosome 3 and, later, chromosomes 6 and 8. [38]. Other groups used FISH on FNAB, 'open biopsies' and enucleation specimens claiming that (a) there was good consistency between the small and larger biopsies, and (b) FISH was reliable for establishing the presence/absence of chromosome 3 and 8 abnormalities, despite some tumour clonal heterogeneity [36, 39, 40]. In a long-term study, however, we demonstrated that FISH lacked sensitivity when only the centromeric probe of chromosome 3 was applied, resulting in small chromosomal abnormalities being missed, and in a false-negative rate of 5% [38]. Other groups also showed the poor sensitivity

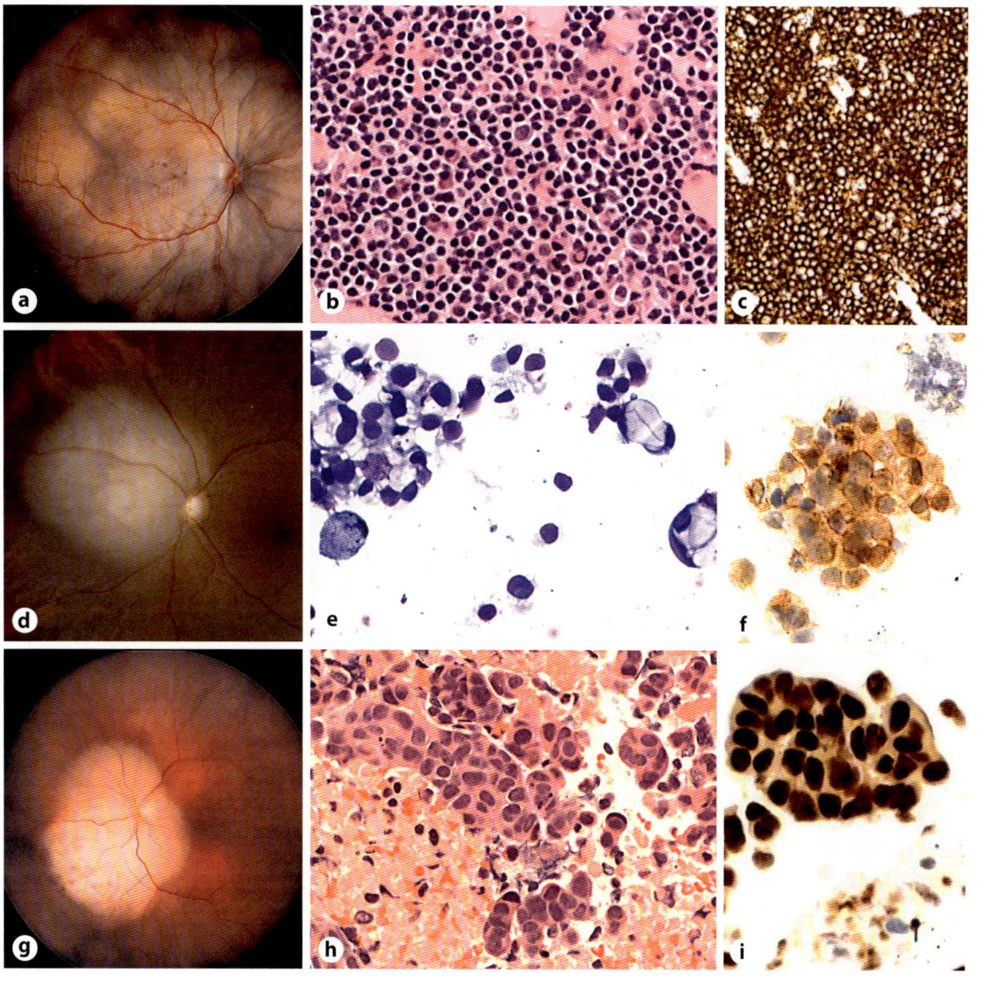

Fig. 2. **a** Funduscopy picture of a 52-year-old male patient with a dense and diffuse creamy orange-coloured choroidal infiltrate. **b** Chorioretinal biopsy revealed a small cell lymphoid infiltrate within the choroid. The cells had a plasmacytic differentiation with occasional scattered Dutcher bodies (HE section, ×20). **c** On immunohistochemical staining, there was clear dominance of the B-cell population over the few admixed reactive T-lymphocytes (PAP staining, ×40). The monotypical nature of the infiltrate could be demonstrated in the IgM stain (not shown). The findings were consistent with a low-grade B-cell lymphoma and it was subtyped as a primary extranodal marginal zone B-cell lymphoma of the choroid. **d** Single nodular peripapillary lesion in the right eye of this 62-year-old man with an otherwise unremarkable medical history. **e** An intraocular biopsy was made and the cytospins demonstrated numerous atypical cells, many of which contained intracytoplasmic vacuoles, positive on mucin stains (MGG, ×60). **f** Immunohistochemistry showed positivity of the cells for cytokeratin and thyroglobulin transcription factor-1, these findings being consistent with a metastatic bronchial carcinoma (PAP stain, ×20). **g** Funduscopy picture in a 48-year-old woman with a previous history of breast carcinoma. **h** Intraocular biopsy produced a cell-rich sample, which was embedded in paraffin. Sections showed numerous cohesive atypical cells with a haemorrhagic and necrotic background (HE section, ×40). **i** Immunostaining for oestrogen receptors revealed positivity of these neoplastic cells, consistent with metastatic breast carcinoma (PAP staining, ×40).

of FISH when compared with molecular genetic techniques, such as microsatellite analysis (MSA) and array comparative genomic hybridization (aCGH) [41, 42]. In 2006, we replaced FISH with MLPA, which examines for genetic gains and losses of 31 loci by means of a multiplex polymerase chain reaction (PCR) [32]. We chose this method as it was affordable by our national health service, because it could be applied to both fresh and formalin-fixed paraffin-embedded tumour material, and it catered for even small intraocular biopsies (fig. 3, 4). We used a kit specifically designed for uveal melanoma [SALSA P027(B1), MRC-Holland, Amsterdam, the Netherlands]. This comprises 12 control probes and test probes directed at 7 loci on chromosome 1, 13 loci on chromosome 3, 6 loci on chromosome 6 and 5 loci on chromosome 8. Prior to its routine implementation, we evaluated MLPA with 73 frozen archival biopsies of choroidal melanomas with long-term clinical follow-up, and found good correlations with survival. We also validated MLPA using both aCGH and MSA in 32 cases of uveal melanoma, demonstrating good concordance between the techniques [Killender et al., in preparation] (fig. 4). Important findings from these studies were that chromosome 3 loss and chromosome 8q gain were associated with increased risk of metastasis even when the MLPA values showed only borderline, or equivocal, abnormality [32]. Such borderline MLPA abnormality was subsequently found to reflect intratumoural melanocytic clonal heterogeneity [23]. Apart from increased sensitivity, MLPA had the additional benefit of requiring smaller samples, thereby making it possible to test small tumour specimens obtained by transscleral FNAB or transretinal biopsy performed with a 25-gauge vitreous cutter. Since 2007, therefore, we have also performed MLPA on biopsies of uveal melanomas treated by radiotherapy or phototherapy. Our recent audit of 452 uveal melanomas examined using MLPA would support the use of MLPA for routine clinical prognostication, especially if the genetic data are considered together with clinical and histological risk factors [33].

Other ocular oncology centres examine chromosomal status in uveal melanoma using aCGH, MSA and high-density single nucleotide polymorphisms (table 2) [43]. For example, Shields et al. [43] reported the feasibility of genetic testing using FNAB material obtained with either a transscleral (52%) or transvitreal (48%) approach in a series of 140 cases at the time of plaque radiotherapy. Adequate DNA was provided for genetic analysis using MSA for monosomy 3 in 75 and 97%, respectively. In a series of 57 eyes treated with iodine-125 plaque brachytherapy, information on chromosome 3 status was obtained in 73% of cases [42]. An RNA-based method, which subdivides into class 1 and class 2 uveal melanomas, is gene expression profiling [44–46]. Genes that discriminate class 1 (low-grade) from class 2 (high-grade) include highly significant clusters of downregulated genes on chromosome 3 and upregulated genes on chromosome 8q, which is consistent with previous cytogenetic studies. This has been adapted into a commercially available PCR-based kit employing 12 signature genes for tumour prediction [47]. Although the above-mentioned techniques vary, they essentially have the same aim, i.e. to identify those uveal melanoma patients with

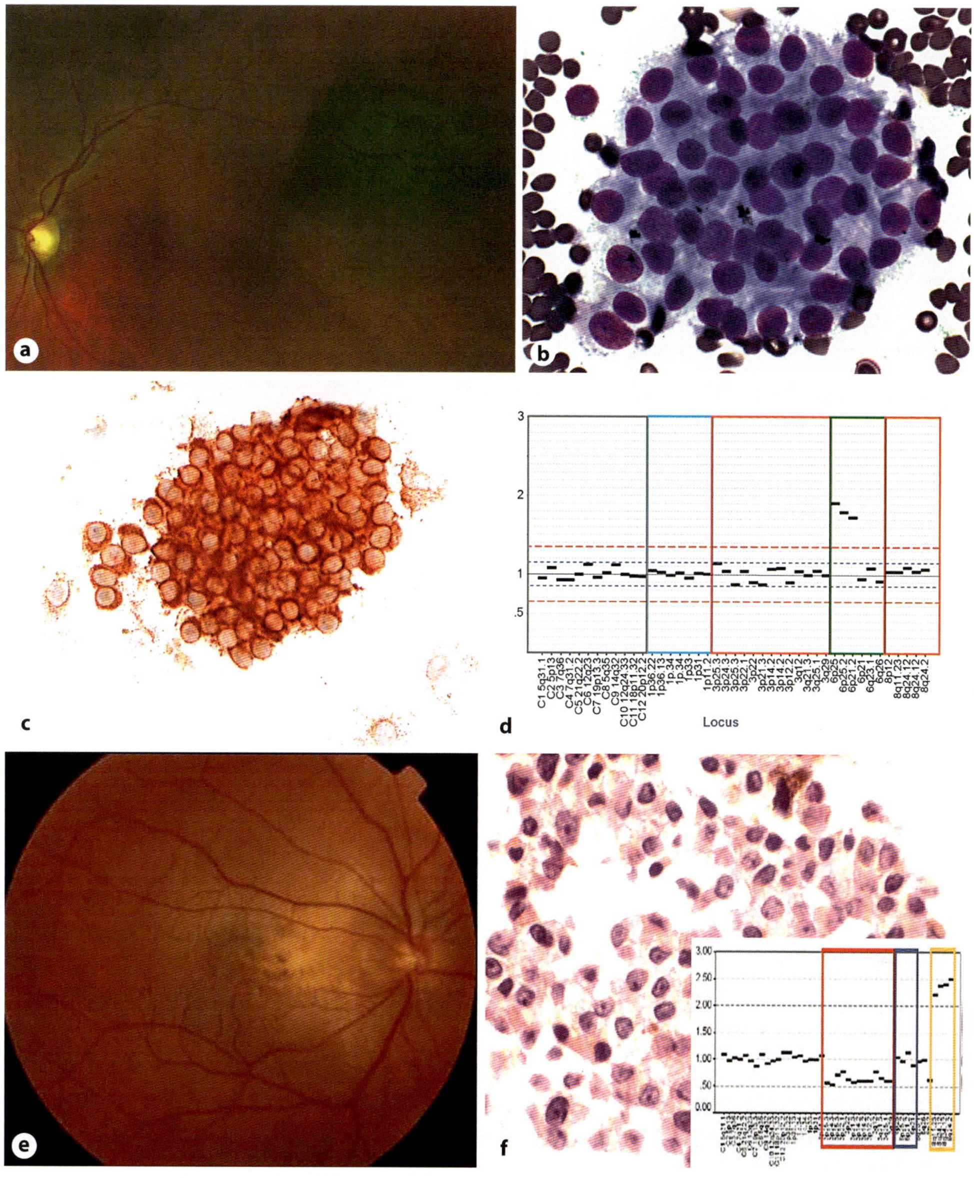

Fig. 3. **a** Funduscopy picture of a 36-year-old male with a choroidal melanocytic lesion of indeterminate nature. **b** The biopsy performed was cellular, comprised of medium-sized cells with moderate pigment-laden cytoplasm, round nuclei and small nucleoli (MGG, ×60). **c** The cells showed immunoreactivity for the melanocytic marker, MelanA (PAP, ×40). **d** DNA extracted from the intraocular biopsy was of sufficient quality and quantity to perform MLPA, examining 31 loci across chromosomes 1, 3, 6 and 8 (y-axis = MLPA dosage quotient). Taken together, the results were consistent with a 'low-grade' choroidal melanoma with a low risk of metastasis. **e** Funduscopy picture of a 48-year-old male with a partially pigmented choroidal tumour. **f** Intraocular biopsy revealed a mixed population of spindle and epithelioid melanoma cells (HE section, ×40). Inset: MLPA demonstrated this tumour to show changes consistent with chromosome 3 loss and gains in 8q (y-axis = MLPA dosage quotient).

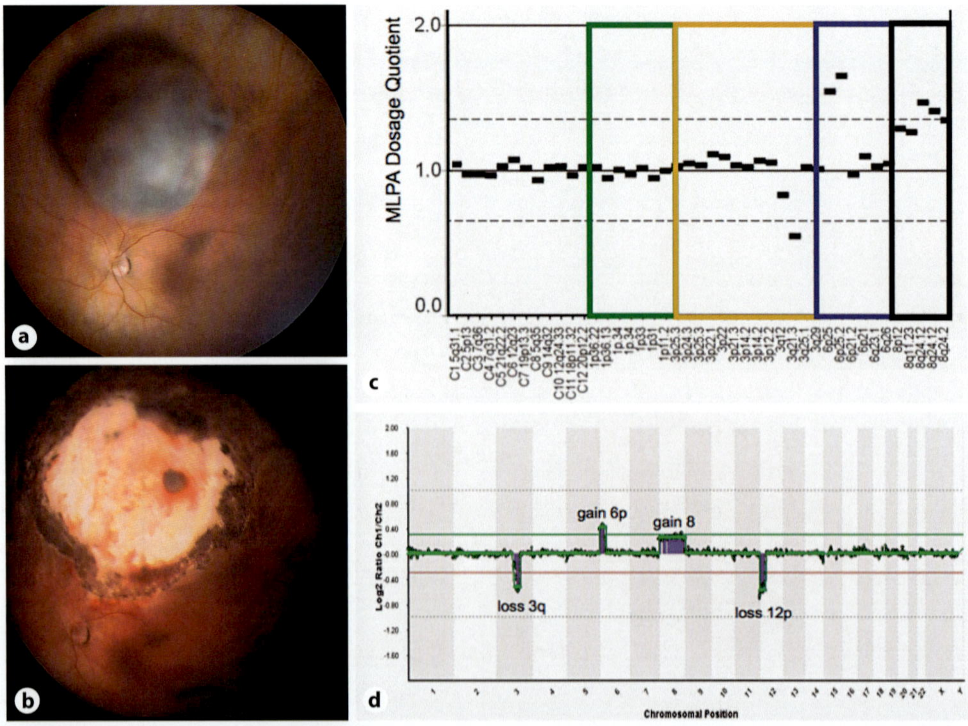

Fig. 4. a Funduscopy of a 55-year-old male with a large pigmented nodular choroidal melanoma. **b** This was treated by endoresection: postoperative picture 3 months after the initial operation. **c** MLPA of the melanoma demonstrated that there were partial deletions in chromosome 3 and increased copies of loci on the short arm of chromosome 6 and on both arms of chromosome 8 (y-axis = MLPA dosage quotient). **d** aCGH of the same melanoma demonstrated similar chromosomal changes, demonstrating good concordance between the two techniques with respect to chromosomes in common. Additional alterations were also seen in chromosome 12 in aCGH (y-axis = log ratio ch1/ch2).

high and low risks of metastasis. The patients are subsequently managed accordingly with respect to counselling, frequency of metastasis screening, and possible inclusion in clinical trials examining adjunctive therapy [48]. A review of questionnaires sent to uveal melanoma patients treated at the Liverpool Ocular Oncology Centre suggests that most patients want to have prognostic analyses performed on their tumours and indeed want to know the results [49].

Examination of Vitreous Samples

Vitreous samples may be acellular (and therefore non-diagnostic in most cases), or cellular. *Acellular* vitreous samples contain any of the following: condensed vitreous strands; iridescent particles such as asteroid hyalosis (calcium soaps) and synchysis scintillans (cholesterol); amyloid deposits; squames, consisting of conjunctival cells artefactually displaced during the vitrectomy procedure; retained lens fragments following cataract removal, and pigment dust. The *cellular* vitreous samples can be categorized as: haemorrhagic, inflammatory non-infectious, inflammatory infectious and neoplastic.

Vitreous specimens with haemorrhage predictably contain varying amounts of erythrocytes, 'ghost cells' and haemosiderin-laden macrophages and acellular eosinophilic material, in various proportions, depending on the age of the haemorrhage. Neoplastic cells or microorganisms within the haemorrhage may reveal the underlying cause.

Inflammatory non-infectious vitritis is characterized by cells consisting predominantly of small T-lymphocytes, which are usually of the CD4+ helper type [20]. Varying proportions of admixed macrophages, monocytes, plasma cells and neutrophils may also be present. Once malignant cells and microorganisms have been excluded, the diagnosis may amount only to a descriptional one, i.e. 'chronic nonspecific vitritis', despite the use of special stains, immunocytology and/or immunoglobulin heavy chain PCR for clonality. In such cases, the clinical history and examination findings are especially important. Inflammatory, infectious vitritis results in vitreous samples composed of abundant cells. Numerous neutrophils are very suggestive of a bacterial (suppurative) endophthalmitis, most commonly caused by organisms such as *Streptococcus* sp., *Staphylococcus aureus, Staphylococcus epidermis,* coagulase-negative staphylococcus, *Neisseria* sp., *Bacillus cereus, Haemophilus influenzae, Propionibacterium acnes, Pseudomonoas aeruginosa, Klebsiella pneumoniae* and *Escerichia coli* [17]. Microbiological cultures or molecular biological techniques are usually necessary for the exact identification of the genus and determination of antibiotic sensitivities [50]; however, bacterial stains of vitreous specimens may identify the causative agent should cultures be negative. It is important to note, however, that neutrophilic infiltrates in the aqueous and/or vitreous can occur with non-bacterial conditions such as Behçet's disease. The presence of eosinophils in the vitreous suggests conditions such as nematode-induced endophthalmitis (e.g. *Toxocara canis*), sympathetic ophthalmia, Lyme disease, and Eales disease. A predominance of macrophages in the vitreous sample may occur in Whipple's disease, ocular toxoplasmosis (fig. 5) [51] as well as endophthalmitis due to *Mycobacterium avium, Histoplasmosis capsulatum, Pneumocystis carinii, Cryptococcus* and *Blastomyces.* In these conditions, fungal stains such as periodic acid-Schiff, Grocott and mucicarmine may reveal cytoplasmic inclusions or cysts of particular sizes and shape (fig. 5). This allows for a morphological suggestion of the possible causative microorganism, to be confirmed, however, with immunocytology, cultures and/or microorganism-directed PCR [52]. Macrophages and multinucleated giant cells suggest a granulomatous process, caused by fungal and mycobacterial infections. The most common fungi causing fungal endophthalmitis include *Candida* sp. (fig. 5), *Aspergillus fumigatus* and *flavus,* as well as *Cryptococcus neoformans.* Acid-fast bacilli, highlighted with Ziehl-Neelsen,

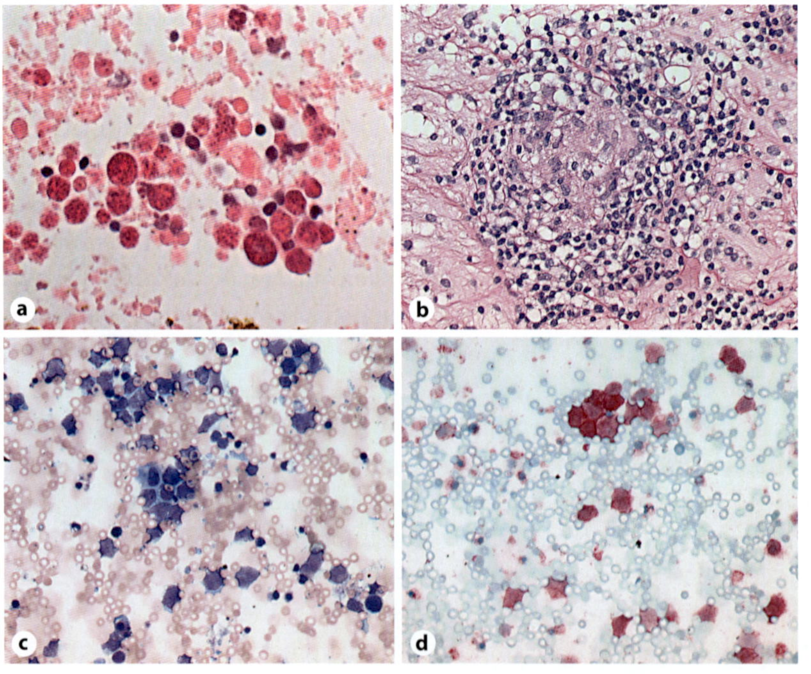

Fig. 5. **a** Macrophage-rich vitreous biopsy in a 73-year-old patient with 'masquerade syndrome' (MGG, ×40). **b** Within some of the macrophages and also lying freely within the biopsy were tachyzoites (arrows), consistent with an ocular toxoplasmosis (MGG, ×60). **c** Multinucleated macrophage with adjacent small macrophages, consistent with a granulomatous vitritis. Included in the differential diagnosis is a mycobacterial infection and sarcoidosis (HE section of an agar block, ×60). **d** Macrophage-rich vitritis caused by *Candida* species in an immunosuppressed patient (MGG, ×40).

are rarely observed in aqueous and vitreous samples but may be found intracytoplasmically within macrophages or retinal pigment epithelial cells in tissue biopsies [53]. In the presence of a granulomatous inflammation and negative staining for fungi, mycobacteria and other microorganisms, the diagnosis of sarcoidosis should be considered, if the history and clinical findings are consistent with this condition. In younger patients, juvenile xanthogranulomatosis should be considered, particularly in anterior uveal lesions (fig. 5). Exceptionally rarely, intranuclear and intracytoplasmic viral inclusion bodies (e.g. cytomegalovirus or herpes simplex) may be demonstrated in association with necrotic cells in the vitreous sample. Such inclusion bodies are, however, more likely to be demonstrated in chorioretinal biopsies, using immunohistochemical or immunofluorescence techniques. PCR analysis of ocular fluid samples may expedite the diagnosis of such conditions, which is important as retinitis can progress rapidly. Syphilis, once the main cause of vitreous opacities [54], now accounts for only about 1% of all vitreous opacities but is increasing in incidence, particularly in the HIV-positive population. It should always be considered in the differential diagnosis of an inflammatory cellular vitreous infiltrate.

Neoplastic disease can be diagnosed in vitreous biopsies with the differential diagnoses of malignant neoplasms including primary ocular tumours such as vitreoretinal lymphoma and retinoblastoma, as well as diseases such as leukaemia, metastatic carcinomas and metastatic cutaneous melanomas (fig. 6) [30]. The main neoplastic 'masquerader' to be diagnosed in a vitrectomy specimen is vitreoretinal lymphoma, which is located most often within the subretinal space, with perivascular infiltrates in the retina and seeding into the vitreous [30]. Cytomorphological examination of vitreous infiltrates in vitreoretinal lymphoma demonstrates medium-to-large cells with minimal cytoplasm, pleomorphic or round (sometimes bare) nuclei and prominent nucleoli (fig. 6). Necrotic material and macrophages are commonly present in the background. Immunocytology shows the neoplastic cells to express B-cell antigens (CD79a, CD20, PAX-5) in the majority of cases (fig. 6). Many of these cells stain positively with the MIB-1 antibody, which indicates a high growth fraction, i.e. a high-grade of malignancy. The diagnosis of lymphoma is supported by demonstration of cellular monoclonality, with monotypic expression of either a light and/or heavy chain of the immunoglobulin gene (usually IgM) [55, 56]. Should sufficient material be available for examination, investigation of the vitreous specimens for rearrangements of the immunoglobulin heavy chain gene using PCR provides further evidence of the neoplastic nature of the lymphocytic infiltrate in vitreoretinal lymphoma. Biochemical analysis of the vitreous specimen for interleukin ratios (IL-10:IL-6) may also support the diagnosis of vitreoretinal lymphoma [57]. If possible, retinal lymphoma should be distinguished from other types of intraocular lymphoma, namely primary uveal lymphoma (fig. 2) and secondary (metastatic) intraocular lymphomas, such as secondary T-cell lymphoma (fig. 6) [30].

Non-Diagnostic Intraocular Biopsy

A 'negative' tumour biopsy is problematic in cases of intraocular malignancy as (a) it fails to provide a diagnosis in a case of suspected neoplasm, and (b) it fails to provide prognostication information to patients whose demands for such information understandably are increasing. The causes of failed intraocular biopsies are many. First of all, the bioptic material may not contain any material of diagnostic relevance, for example, vitreous in primary uveal lymphomas, and in some cases of retinal lymphoma where the tumour cells are 'hidden' in the subretinal space with minimal vitreal involvement. Second, the patient may have been treated with medications (e.g. steroids) prior to intraocular biopsy (e.g. vitrectomy), increasing tumour cell fragility. Thirdly, the size of the biopsy sample may be insufficient, containing only scanty tumour cells compared to the total cell population. Fourthly, the specimen may not be handled properly, for example, not being placed in the correct fixative or being left unfixed for an excessive time. Fifthly, the specimen may be lost during laboratory processing (e.g. a tiny retinal biopsy), perhaps as a result of a technical error. Most of

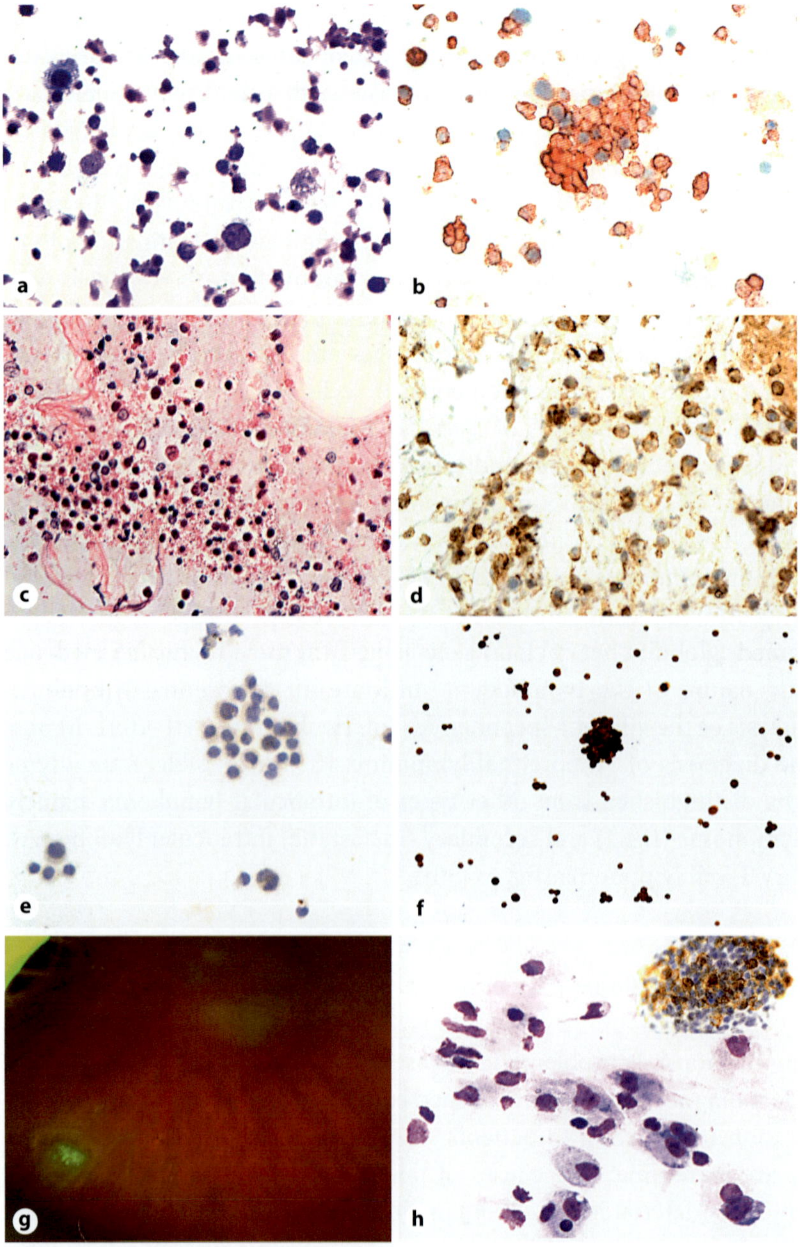

Fig. 6. **a** Moderately cell dense vitrectomy specimen composed of numerous lymphoid blasts with admixed macrophages and lytic cells in the background (MGG, ×40). **b** Immunoreactivity of the lymphoid blasts for the B-cell antigen CD20, consistent with a high-grade vitreoretinal lymphoma (diffuse large B-cell lymphoma) (APAAP, ×40). **c** Dense cellular infiltrate of a vitrectomy specimen processed in paraffin. The cells are small to medium in size with a narrow cytoplasmic rim, large nuclei and dense nuclear chromatin. In the background are lytic cells and macrophages (HE, ×20). **d** The neoplastic cells are immunoreactive for CD2, CD3 and CD8, consistent with a intraocular manifestation of an intestinal T-cell lymphoma (enteropathy-associated T-cell lymphoma; (PAP, ×40). **e** Vitrectomy specimen from a 50-year-old patient with a previous history of cutaneous melanoma. Scattered and aggregates of tumour cells, with a rim of cytoplasm containing melanin pigment, are

these problems can be avoided by taking the precautions already mentioned in this article, and by working closely with an experienced biomedical scientist and reporting pathologist. Whenever a non-diagnostic biopsy occurs, it is especially important for the clinician and pathologist to confer without delay, so that any errors can be avoided if the investigation is repeated.

Conclusions

In agreement with others, our experience would suggest that biopsy of intraocular malignancies is a valuable procedure for both diagnostic and prognostic purposes in adults. In experienced hands, short-term local complications are rare. Harvesting tissue increases the probability of achieving a correct and unequivocal diagnosis, and is often the only way to establish a rare diagnosis. The limitations of an intraocular biopsy procedure should be taken seriously: for example, an FNAB is not always a reliable diagnostic tool with choroidal lesions thinner than 2 mm. Additional immunohistochemical and molecular pathological examinations will probably result in most biopsy procedures being performed in major centres, where there is expertise in their preparation and interpretation by ocular pathologists. At present, molecular genetic techniques combined with clinical and histomorphological data allow for individualized prognostication and management plans for the patient. We envisage in the future that intraocular biopsy will be used to test primary tumours for markers in determining treatment in ocular oncology, as happens with the HER2 marker in breast cancer cells. Targeted therapy with drugs designed for specific molecular pathways could then be administered at initial diagnosis according to the particular melanoma subtype. In this way, the present grim prognosis of metastatic uveal melanoma might be improved.

seen. **f** Immunoreactivity of the pigment-containing cells for MelanA, consistent with an intraocular metastasis of the skin melanoma (PAP, ×20). **g** Funduscopy picture of a 69-year-old patient with a 1-year history of acute myeloid leukaemia and a significant vitreous infiltrate. **h** Vitreous biopsy demonstrated numerous medium-sized cells with moderate cells containing basophilic granula (MGG, ×40). Inset: these cells were positive for myeloperoxidase (PAP, ×20). Taken together, the findings were consistent with an intraocular leukaemic manifestation.

Acknowledgements

I am deeply indebted to Prof. Bertil Damato, Head of the Liverpool Ocular Oncology Centre, who has critically read and made improvements to the manuscript before its submission. The author would also like to thank Prof. Heinrich Heimann and Mr. Carl Groenwald (Consultant Ophthalmologists at the Royal Liverpool University Hospital) for the provision of intraocular biopsies. All of the clinical photographs included in this chapter were provided by the clinical photographic archives of the Liverpool Ocular Oncology Centre and have been included for publication with patient consent.

References

1 Char DH, Kroll SM, Stoloff A, Kaleta-Michaels S, Crawford JB, Miller TR, Howes EL Jr, Ljung BM: Cytomorphometry of uveal melanoma. Comparison of fine needle aspiration biopsy samples with histologic sections. Anal Quant Cytol Histol 1991;13:293–299.

2 Char DH, Miller T: Accuracy of presumed uveal melanoma diagnosis before alternative therapy. Br J Ophthalmol 1995;79:692–696.

3 Jensen OA, Andersen SR: Late complications of biopsy in intraocular tumors. Acta Ophthalmol (Copenh) 1959;37:568–575.

4 Midena E, Segato T, Piermarocchi S, Boccato P: Fine needle aspiration biopsy in ophthalmology. Surv Ophthalmol 1985;29:410–422.

5 Damato BE, Foulds WS: Ciliary body tumours and their management. Trans Ophthalmol Soc UK 1986; 105:257–264.

6 Char DH, Miller TR, Ljung BM, Howes EL Jr, Stoloff A: Fine needle aspiration biopsy in uveal melanoma. Acta Cytol 1989;33:599–605.

7 Foulds WS: The uses and limitations of intraocular biopsy. Eye (Lond) 1992;6:11–27.

8 Shields JA, Shields CL, Ehya H, Eagle RC Jr, De Potter P: Fine-needle aspiration biopsy of suspected intraocular tumors. The 1992 Urwick Lecture. Ophthalmology 1993;100:1677–1684.

9 Augsburger JJ: Diagnostic biopsy of selected intraocular tumors. Am J Ophthalmol 2005;140: 1094–1095.

10 Damato B, Jones AG: Uveal melanoma: resection techniques. Ophthalmol Clin North Am 2005;18: 119–128.

11 Cole CJ, Kwan AS, Laidlaw DA, Aylward GW: A new technique of combined retinal and choroidal biopsy. Br J Ophthalmol 2008;92:1357–1360.

12 Coupland SE: The pathologist's perspective on vitreous opacities. Eye (Lond) 2008;22:1318–1329.

13 Raparia K, Chang CC, Chevez-Barrios P: Intraocular lymphoma: diagnostic approach and immunophenotypic findings in vitrectomy specimens. Arch Pathol Lab Med 2009;133:1233–1237.

14 Eide N, Walaas L: Fine-needle aspiration biopsy and other biopsies in suspected intraocular malignant disease: a review. Acta Ophthalmol 2009;87:588–601.

15 Coupland SE, Perez-Canto A, Hummel M, Stein H, Heimann H: Assessment of HOPE fixation in vitrectomy specimens in patients with chronic bilateral uveitis (masquerade syndrome). Graefes Arch Clin Exp Ophthalmol 2005;243:847–852.

16 Wang HH, Sovie S, Trawinski G, Garcia LW, Abu-Jawdeh GM, Upton M, Werneke S: ThinPrep processing of endoscopic brushing specimens. Am J Clin Pathol 1996;105:163–167.

17 Green WR: Diagnostic cytopathology of ocular fluid specimens. Ophthalmology 1984;91:726–749.

18 Mandell DB, Levy JJ, Rosenthal DL: Preparation and cytologic evaluation of intraocular fluids. Acta Cytol 1987;31:150–158.

19 Augsburger JJ: Invasive diagnostic techniques for uveitis and simulating conditions. Trans Pa Acad Ophthalmol Otolaryngol 1990;42:964–971.

20 Davis JL, Solomon D, Nussenblatt RB, Palestine AG, Chan CC: Immunocytochemical staining of vitreous cells. Indications, techniques, and results. Ophthalmology 1992;99:250–256.

21 Obata H, Horiuchi H, Tsuru T: Clinical application of new membrane filter for cytopathological diagnosis in ophthalmology. Jpn J Ophthalmol 1993;37: 344–351.

22 Coupland SE, Bechrakis NE, Anastassiou G, Foerster AM, Heiligenhaus A, Pleyer U, Hummel M, Stein H: Evaluation of vitrectomy specimens and chorioretinal biopsies in the diagnosis of primary intraocular lymphoma in patients with Masquerade syndrome. Graefes Arch Clin Exp Ophthalmol 2003;241:860–870.

23 Dopierala J, Damato BE, Lake SL, Taktak AF, Coupland SE: Genetic heterogeneity in uveal melanoma assessed by multiplex ligation-dependent probe amplification. Invest Ophthalmol Vis Sci 2010;51:4898–4905.
24 Shields CL, Manquez ME, Ehya H, Mashayekhi A, Danzig CJ, Shields JA: Fine-needle aspiration biopsy of iris tumors in 100 consecutive cases: technique and complications. Ophthalmology 2006;113:2080–2086.
25 Char DH, Crawford JB, Kroll S: Iris melanomas. Diagnostic problems. Ophthalmology 1996;103:251–255.
26 Schalenbourg A, Coupland S, Kacperek A, Damato B: Iridocyclectomy for neovascular glaucoma caused by proton-beam radiotherapy of pigmented ciliary adenocarcinoma. Graefes Arch Clin Exp Ophthalmol 2008;246:1499–1501.
27 Tiberti A, Damato B, Hiscott P, Vora J: Iris ectopic thyroid tissue: report of a case, Arch Ophthalmol 2006;124:1497–1500.
28 Cohen VM, Dinakaran S, Parsons MA, Rennie IG: Transvitreal fine needle aspiration biopsy: the influence of intraocular lesion size on diagnostic biopsy result. Eye (Lond) 2001;15:143–147.
29 Augsburger JJ, Correa ZM, Schneider S, Yassin RS, Robinson-Smith T, Ehya H, Trichopoulos N: Diagnostic transvitreal fine-needle aspiration biopsy of small melanocytic choroidal tumors in nevus versus melanoma category. Trans Am Ophthalmol Soc 2002;100:225–232.
30 Coupland SE, Damato B: Understanding intraocular lymphomas. Clin Experiment Ophthalmol 2008; 36:564–578.
31 Coupland SE, Vorum H, Mandal N, Kalirai H, Honore B, Urbak SF, Lake SL, Dopierala J, Damato B: Proteomics of uveal melanomas suggests HSP-27 as a possible surrogate marker of chromosome 3 loss. Invest Ophthalmol Vis Sci 2010;51:12–20.
32 Damato B, Dopierala J, Klaasen A, van Dijk M, Sibbring J, Coupland SE: Multiplex ligation-dependent probe amplification of uveal melanoma: correlation with metastatic death. Invest Ophthalmol Vis Sci 2009;50:3048–3055.
33 Damato BE, Dopierala JA, Coupland SE: Genotypic profiling of 452 choroidal melanomas with multiplex ligation-dependent probe amplification. Clin Cancer Res 2010;16:6083–6092.
34 Damato B, Eleuteri A, Fisher AC, Coupland SE, Taktak AF: Artificial neural networks estimating survival probability after treatment of choroidal melanoma. Ophthalmology 2008;115:1598–1607.
35 Prescher G, Bornfeld N, Becher R: Nonrandom chromosomal abnormalities in primary uveal melanoma. J Natl Cancer Inst 1990;82:1765–1769.
36 Sisley K, Nichols C, Parsons MA, Farr R, Rees RC, Rennie IG: Clinical applications of chromosome analysis, from fine needle aspiration biopsies, of posterior uveal melanomas. Eye (Lond) 1998;12: 203–207.
37 White VA, Chambers JD, Courtright PD, Chang WY, Horsman DE: Correlation of cytogenetic abnormalities with the outcome of patients with uveal melanoma. Cancer 1998;83:354–359.
38 Damato B, Duke C, Coupland SE, Hiscott P, Smith PA, Campbell I, Douglas A, Howard P: Cytogenetics of uveal melanoma: a 7-year clinical experience. Ophthalmology 2007;114:1925–1931.
39 Midena E, Bonaldi L, Parrozzani R, Radin PP, Boccassini B, Vujosevic S: In vivo monosomy 3 detection of posterior uveal melanoma: 3-year follow-up. Graefes Arch Clin Exp Ophthalmol 2008; 246:609–614.
40 Maat W, Jordanova ES, van Zelderen-Bhola SL, Barthen ER, Wessels HW, Schalij-Delfos NE, Jager MJ: The heterogeneous distribution of monosomy 3 in uveal melanomas: implications for prognostication based on fine-needle aspiration biopsies. Arch Pathol Lab Med 2007;131:91–96.
41 Onken MD, Worley LA, Person E, Char DH, Bowcock AM, Harbour JW: Loss of heterozygosity of chromosome 3 detected with single nucleotide polymorphisms is superior to monosomy 3 for predicting metastasis in uveal melanoma. Clin Cancer Res 2007;13:2923–2927.
42 Young TA, Burgess BL, Rao NP, Gorin MB, Straatsma BR: High-density genome array is superior to fluorescence in-situ hybridization analysis of monosomy 3 in choroidal melanoma fine needle aspiration biopsy. Mol Vis 2007;13:2328–2333.
43 Shields CL, Ganguly A, Bianciotto CG, Turaka K, Tavallali A, Shields JA: Prognosis of uveal melanoma in 500 cases using genetic testing of fine-needle aspiration biopsy specimens. Ophthalmology 2011;118:396–401.
44 Onken MD, Worley LA, Davila RM, Char DH, Harbour JW: Prognostic testing in uveal melanoma by transcriptomic profiling of fine needle biopsy specimens. J Mol Diagn 2006;8:567–573.
45 Onken MD, Worley LA, Ehlers JP, Harbour JW: Gene expression profiling in uveal melanoma reveals two molecular classes and predicts metastatic death. Cancer Res 2004;64:7205–7209.
46 Harbour JW: Molecular prognostic testing in uveal melanoma: has it finally come of age? Arch Ophthalmol 2007;125:1122–1123.
47 Onken MD, Worley LA, Harbour JW: Association between gene expression profile, proliferation and metastasis in uveal melanoma. Curr Eye Res 2010; 35:857–863.

48 Damato B, Coupland SE: Translating uveal melanoma cytogenetics into clinical care. Arch Ophthalmol 2009;127:423–429.
49 Cook SA, Damato B, Marshall E, Salmon P: Psychological aspects of cytogenetic testing of uveal melanoma: preliminary findings and directions for future research. Eye (Lond) 2009;23:581–585.
50 Adra CN, Eggerding FA: Diagnostic molecular microbiology in posterior uveitis: principles and applications. Int Ophthalmol Clin 1995;35:75–92.
51 Palkovacs EM, Correa Z, Augsburger JJ, Eagle RC Jr: Acquired toxoplasmic retinitis in an immunosuppressed patient: diagnosis by transvitreal fine-needle aspiration biopsy. Graefes Arch Clin Exp Ophthalmol 2008;246:1495–1497.
52 Fardeau C, Romand S, Rao NA, Cassoux N, Bettembourg O, Thulliez P, Lehoang P: Diagnosis of toxoplasmic retinochoroiditis with atypical clinical features. Am J Ophthalmol 2002;134:196–203.
53 Rao NA, Saraswathy S, Smith RE: Tuberculous uveitis: distribution of Mycobacterium tuberculosis in the retinal pigment epithelium. Arch Ophthalmol 2006;124:1777–1779.
54 Duke-Elder S: Diseases of the Vitreous Body. London, Kimpton, 1969.
55 Coupland SE, Heimann H, Bechrakis NE: Primary intraocular lymphoma: a review of the clinical, histopathological and molecular biological features. Graefes Arch Clin Exp Ophthalmol 2004;242:901–913.
56 Chan CC: Molecular pathology of primary intraocular lymphoma. Trans Am Ophthalmol Soc 2003;101:275–292.
57 Merle-Beral H, Davi F, Cassoux N, Baudet S, Colin C, Gourdet T, Bodaghi B, LeHoang P: Biological diagnosis of primary intraocular lymphoma. Br J Haematol 2004;124:469–473.

Prof. Sarah E. Coupland, MBBS, PhD, FRCPath
Department of Cellular and Molecular Pathology
University of Liverpool, 5th Floor Duncan Building
Daulby Street, Liverpool L69 3GA (UK)
E-Mail s.e.coupland@liverpool.ac.uk

Jager MJ, Desjardins L, Kivelä T, Damato BE (eds): Current Concepts in Uveal Melanoma.
Dev Ophthalmol. Basel, Karger, 2012, vol 49, pp 117–136

Anti-Angiogenic Therapy in Uveal Melanoma

Mariam el Filali · Pieter A. van der Velden · Gregorius P.M. Luyten · Martine J. Jager

Department of Ophthalmology, Leiden University Medical Centre (LUMC), Leiden, The Netherlands

Abstract

For several decades, targeting of tumor-related vessels has been regarded as a potential anticancer therapy. Such anti-angiogenic therapy is based on the assumption that a tumor cannot grow beyond the limits of diffusion (about 1–2 mm) of oxygen and nutrients from capillaries, unless angiogenesis takes place. Vascular endothelial growth factor (VEGF) plays a key role in angiogenesis, regulating vasopermeability as well as the proliferation and migration of endothelial cells. In several types of cancer (colon carcinoma, soft tissue sarcomas and gastric cancer), serum VEGF levels are a marker for disease stage and an indicator of metastasis. VEGF levels are significantly elevated in uveal melanoma patients with metastatic disease compared to patients without metastases. Anti-angiogenic therapy, such as bevacizumab, is currently used for the treatment of metastases of several malignancies. Anti-angiogenic therapy has not yet been tested for the treatment of primary uveal melanoma or related metastatic disease. Clinicians, however, have a broad experience with anti-angiogenic agents in patients with uveal melanoma by treating the complications of radiation therapy. We will discuss tumor angiogenic processes and related molecular pathways in uveal melanoma. The role of VEGF and the potential use of current commercially and experimentally available anti-angiogenic drugs for the treatment of primary uveal melanoma and/or metastatic disease will be explained below.

The targeting of tumor-related vessels has been investigated for several decades. The fundamental belief that a tumor cannot grow beyond the limits of diffusion (about 1–2 mm) of oxygen and nutrients from blood vessels has been advocated since the 1970s. Dr. J. Folkman has played a pivotal role, describing the molecular aspects of tumor angiogenesis and also predicting anti-angiogenic therapy. Furthermore, he demonstrated the importance of a potent tumor blood supply for the growth of metastases [1]. In uveal melanoma, metastasis occurs exclusively via the hematogenous route, so that the tumor vasculature is important [2].

Criscuolo et al. [3] were the first to describe the occurrence of vascular permeability increasing factor in malignant glioma, which we currently know as vascular endothelial growth factor (VEGF). The VEGF-A isotype, referred to as VEGF in this review, plays a key role in angiogenesis, regulating vasopermeability as well as the proliferation and migration of endothelial cells [4]. In several tumors (e.g. colon carcinoma, soft tissue sarcomas and gastric cancer), serum VEGF levels have been found to be a marker of disease stage and an indicator of metastasis [5–7]. In uveal melanoma, VEGF expression in serum of patient with a primary tumor, cannot predict survival. Nevertheless, serum VEGF levels are significantly higher in uveal melanoma patients with metastatic disease compared to patients without such spread [8, 9].

In 2004, bevacizumab, the first angiogenesis inhibitor that targets VEGF, was developed, approved and licensed for intravenous infusion in the treatment of colorectal carcinoma [10]. Bevacizumab is also used for the treatment of metastases from several other malignancies, including renal and lung cancer [11, 12] and is under investigation for other primary tumors (e.g. pancreas cancer and cutaneous melanoma) [13, 14]. However, not all results are positive. It has been reported that VEGF inhibitors elicit tumor adaptation and increased lymphatic and distant metastasis in patients with pancreatic neuroendocrine carcinoma and glioblastoma-bearing mice [15]. In uveal melanoma, antiangiogenesis therapy has not yet been used for the treatment of primary uveal melanoma or related metastatic disease. Still, there has been extensive research into the effect of anti-angiogenic agents such as bevacizumab on uveal melanoma cells and animal models [16; el Filali et al., submitted]. Serendipitously, clinicians already have a broad experience with anti-angiogenic agents in patients with uveal melanoma as a result of treating the complications of radiation therapy of the primary tumor.

The role of VEGF as key mediator in tumor angiogenesis and as a main treatment target will be addressed as well as several other anti-angiogenic drugs for future treatment of primary uveal melanoma and/or metastatic disease.

Tumor Angiogenesis in Uveal Melanoma

Several authors have investigated the role of blood vessels in uveal melanoma growth and metastasis.

Vascular Density

Microvessel density of uveal melanoma was studied by immunohistochemistry and has been found to correlate strongly with the risk of metastatic death [17]. Microvessel density was shown to be locally induced and not evenly distributed in the whole tumor [18, 19]. In subsequent studies, specific 'hot spots' of vascular density have been shown to correlate with uveal melanoma-related metastatic death [20].

Extracellular Matrix Patterns and Vasculogenic Mimicry

Several extracellular matrix patterns have been described in uveal melanoma. When so-called closed loops and networks are present, they predict a worse 10-year probability of melanoma-specific survival (loops: 0.45 vs. 0.83; two-sided $p < 0.0001$, and networks: 0.41 vs. 0.72, two-sided $p < 0.0001$) [21]. Moreover, these patterns are also shown in uveal melanoma-related metastases and are described as 'vasculogenic mimicry'. This concept proposes the formation of fluid-conducting channels by tumor cells independent of local vascular outgrowth, without endothelium [22]. Vasculogenic mimicry has also been identified in several other malignancies and shown to be associated with aggressive tumor behavior [23]. The increased diffusion surface that these channels offer could allow continued growth of uveal melanoma.

Vasculature and Metastatic Disease

For metastases to occur, uveal melanoma cells must detach from the primary tumor and invade surrounding tissues to enter a nearby blood vessel, after which the cell can circulate systemically to a new location. A strong association has been observed between tumor cell ingrowth into blood vessels and extraocular extension, which is known to indicate a poor survival probability [24, 25]. To form a metastatic tumor, the circulating malignant cells must exit the circulation and enter an organ, which in case of uveal melanoma is usually the liver [26, 27]. The predominance of liver metastasis cannot be explained solely by blood circulation because the lungs are the first organ that uveal melanoma cells encounter. There must be a preferential microenvironment in which uveal melanoma cells proliferate more easily or quickly. Expression of insulin growth factor-1 receptor (IGF-1R) in uveal melanoma offers a possible explanation for the bad prognosis [28]. IGF-1, the ligand for IGF-1R, leads to phosphorylation of IGF-1R, which in turn activates key signal molecules involved in cell proliferation [29]. IGF-1 is mainly produced by the liver and may explain the preferential growth of hepatic metastasis from uveal melanoma [29]. Besides a favorable microenvironment, the new location must provide a good blood supply. Interestingly, IGF-1 has been shown to stimulate secretion of VEGF in retinal pigment epithelial cells and possibly IGF-1 signaling is also involved in tumor angiogenesis in hepatic metastases from uveal melanoma [30].

Molecular Mediators of Angiogenesis

Uveal melanoma is characterized by slow progression and periods of dormancy, both of the primary tumor and of metastases. It has been suggested that this dormancy is associated with an avascular phase, in which a conversion to the angiogenic phenotype has yet to be established. This conversion, which is known as the 'angiogenic switch', is due to an alteration in the balance of inhibitory and stimulatory factors [31]. Folkman hypothesized that one important stimulatory factor, called tumor angiogenic factor, induces the tumor to convert to such an angiogenic phenotype. VEGF was later identified as one of the most potent tumor angiogenic factor molecules, which acts as the central mediator of tumor angiogenesis by regulating vasopermeability and the proliferation and migration of endothelial cells [4].

Another group of enzymes that has been implicated in tumor angiogenesis and the associated tissue remodeling is the family of metalloproteinases (MMPs) [32]. The major MMPs involved in tumor angiogenesis are MMP-2, -9, and -14 [33]. The survival rate of patients with MMP-2- and MMP-9-positive uveal melanomas is worse than that of patients with MMP-2- and MMP-9-negative melanomas (31–27 vs. 85%, $p < 0.05$) [34]. Epidermal growth factor (EGF) and its receptor (EGFR) also have an established role in tumorigenesis. EGF(R) is a potent proangiogenic factor able to induce migration of endothelial cells and regulate the production of angiogenic factors in tumor cells, such as VEGF, basic fibroblast growth factor and angiopoietin [35]. Other common angiogenic factors detected in tumors are platelet-derived growth factor, hepatocyte growth factor, and IGF family members [36]. In uveal melanoma, several of these proangiogenic factors have been analyzed. Boyd et al. [37], for example, demonstrated uveal melanoma cell expression of basic fibroblast growth factor at the protein level by immunohistochemistry and by RT-PCR in almost all tested samples (89%), especially around microvasculature. Expression of the receptors for hepatocyte growth factor and IGF are bad prognostic factors in uveal melanoma as described earlier [28].

The Role of Vascular Endothelial Growth Factor

Structure

VEGF-α, also termed VEGF-A or VEGF, is a member of the VEGF platelet-derived growth factor family that also comprises placenta growth factor [38]. VEGF exists in a range of isoforms due to alternative splicing of the RNA: $VEGF_{121}$, $VEGF_{165}$ (the predominant form), $VEGF_{189}$, and $VEGF_{206}$ [39]. VEGF proteins are available to cells by at least two different mechanisms: (1) as freely diffusible proteins ($VEGF_{121}$, $VEGF_{165}$), or (2) after protease activation and cleavage of protein bound to heparin ($VEGF_{189}$, $VEGF_{206}$) [40].

The effects of VEGF are mainly mediated through binding to VEGF receptor 1 (Flt-1) and VEGF receptor 2 (KDR), both of which are expressed on vascular endothelial cells as well as on tumor cells and on other cells in the tumor microenvironment [41]. Flt-1 and KDR are transmembrane tyrosine kinase receptors that become active upon ligand binding and thereby trigger signal transduction pathways that are involved in angiogenesis. VEGF receptor 3 (Flt-4) is mainly involved in VEGF-C- and VEGF-D-mediated lymphangiogenesis [42].

Genetics

The human VEGF gene has been assigned to chromosome 6p21.3 [43]. Chromosome 6p gain in uveal melanoma has been reported in several studies and includes the VEGF locus [44, 45]. A correlation between copy number changes in the 6p region and the expression of VEGF in uveal melanoma has not yet been established [46]. Moreover, abnormalities of chromosome 6 have been associated with a better survival in uveal melanoma patients, which seems to be in conflict with metastasis-promoting tumor angiogenesis [47]. Unfortunately, in this study, loss of 6q and gain of 6p were combined as one factor, thus limiting the evaluation of the role of 6p unclear.

Regulation

Several factors have been shown to participate in the regulation of VEGF expression. However, hypoxia is the best known factor and VEGF mRNA expression can be induced reversibly by exposure to low oxygen levels in many cell types [48]. The key regulator of hypoxia-induced VEGF is the transcription factor hypoxia-inducible factor (HIF)-1α [49]. Under hypoxic conditions, HIF-1α is stabilized and drives the expression of a large cluster of genes including VEGF and erythropoietin [50]. In tumors with significant necrosis, the expression of VEGF is mostly upregulated in the ischemic tumor cells adjacent to the necrotic areas [51].

Several cytokines or growth factors, such as EGF, platelet-derived growth factor, transforming growth factor β, interleukin 6, interleukin 1 and IGF-1, are also known to upregulate VEGF expression in several normal cells, including retinal pigment epithelial cells, and in tumor cells [4, 30, 52]. Uveal melanomas that overexpress one of these cytokines/growth factors or the receptors for these ligands might generate autocrine signaling that promotes tumor growth and tumor vascularization. Blocking IGFR with picropodophyllin in mice with induced choroidal neovascularization reduced VEGF levels and vessel formation [53]. In addition, picropodophyllin has been shown to inhibit uveal melanoma growth in vivo in uveal melanoma xenografts [54]. Furthermore, VEGF expression has been demonstrated to be increased in association with specific genetic events such as loss of tumor suppressor genes or activation of oncogenes. The von Hippel-Lindau tumor suppressor gene has been implicated in the regulation of VEGF gene expression [55]. Loss of von Hippel-Lindau protein function results in constitutive activation of HIF-1α and thus VEGF expression [56, 57].

Oncogenic mutations or amplification of *ras* and overexpression of v-*Src* have also been shown to upregulate VEGF [58]. Interestingly, we have demonstrated high Src activation in uveal melanoma that is associated with a constitutive activation of the mitogen-activated protein kinase (MAPK) pathway and correlated with a bad prognosis [59; el Filali et al., submitted].

Biological Function

VEGF is known to be involved in several different aspects of angiogenesis. After binding of VEGF to the VEGFR-1 and -2, several proteins are activated including focal adhesion kinase, PI3K and Src. These downstream kinases promote vascular permeability, endothelial cell proliferation, migration and survival [60]. Originally, VEGF was referred to as vascular permeability factor [61]. A rapid increase in vascular permeability occurs when the microvasculature is exposed acutely to any number of vascular permeabilizing factors, like VEGF, allowing the diffusion of trophic substances to adjacent tumor cells. VEGF promotes proliferation of endothelial cells through induction of the Raf-MEK-MAPK pathway and the formation of the endothelial lining of tumor vessels by attracting circulating endothelial cells. VEGF also activates focal adhesion kinase and the PI3-kinase-Akt pathway, inducing subsequent migration of endothelial cells expressing VEGFR-2 [60]. In addition, VEGF is involved in cell survival (via PI3-kinase/Akt activation and antiapoptotic proteins) and monocyte activation, the description of which is beyond the scope of this chapter [62, 63].

Expression and Implication of Vascular Endothelial Growth Factor in Uveal Melanoma

VEGF induction has been extensively demonstrated in a range of malignancies, including lung, breast, and gastrointestinal tract tumors [64–66]. In the eye, VEGF gene and protein expression are observed in ocular tissues, primarily in the retina and retinal pigment epithelium, and are particularly upregulated in retinopathies that are associated with angiogenic proliferation [67]. The first study that investigated VEGF gene expression in uveal melanoma applied RT-PCR to 7 uveal melanoma cell lines [68]. Subsequently, Sheidow et al. [69] showed VEGF immunostaining in uveal melanoma samples of enucleated eyes, but did not find any correlation between the occurrence of metastatic disease and the amount of VEGF expression in uveal melanoma tissue.

Using immunohistochemistry, Boyd et al. [70] showed only a moderate staining of VEGF in uveal melanoma samples (22%; n = 50). On the contrary, all uveal melanomas tested expressed VEGF mRNA (n = 20). Another publication by the same investigators describes elevated VEGF concentrations (up to 21.6 ng/ml) in vitreous and anterior chamber fluids of eyes with uveal melanoma compared to samples from healthy eyes (<0.96 ng/ml). Remarkably, the highest VEGF levels were found in fluids of eyes that had been treated with radiation [37]. In studies by Missotten et al. [71] and others, elevated VEGF in the aqueous humor of eyes with uveal melanoma was

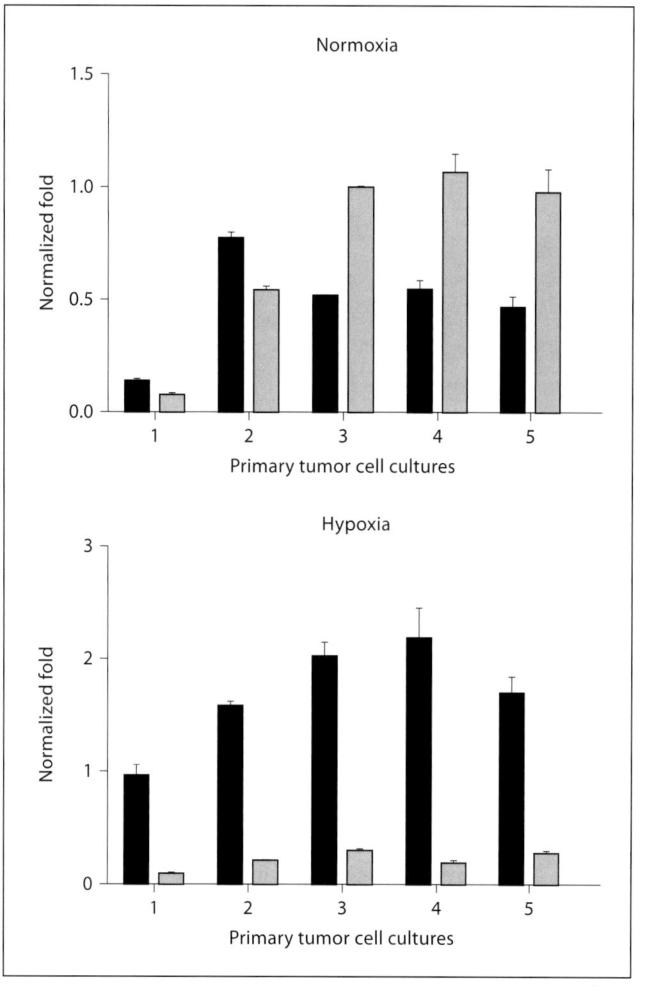

Fig. 1. VEGF and TSP-1 mRNA expression in primary uveal melanoma cell cultures. The amount of VEGF (black) and TSP-1 (gray) mRNA expression was measured with quantitative real-time RT-PCR in primary uveal melanoma cell cultures (cultures 1–5) under normoxic (1% O_2) and hypoxic (20% O_2) conditions after 24 h. Expression is demonstrated in normalized fold.

confirmed and found to be correlated with largest basal tumor diameter and tumor height. In situ analysis further demonstrated that both the tumor cells as well as the retina cells express VEGF [71]. We further investigated the regulation of VEGF in uveal melanoma and found that hypoxia massively induces HIF-1α and VEGF in uveal melanoma cell lines and primary tumor cell cultures. On the contrary and as expected, TSP-1, an anti-angiogenic factor, was downregulated when uveal melanoma cells were exposed to ischemic conditions (fig. 1). VEGF expression in primary uveal melanoma samples (n = 27) was variable (range of expression 0.04–9.55 normalized fold), and demonstrated no correlation with specific histological markers or prognosis. Upregulation of VEGF in uveal melanoma cell lines, in response to hypoxia, did not increase cell proliferation [9]. The ability to modulate expression of VEGF by uveal melanoma cells may provide the tumor with the opportunity to initiate vascularization. Whether VEGF is essential for tumor angiogenesis should be analyzed in

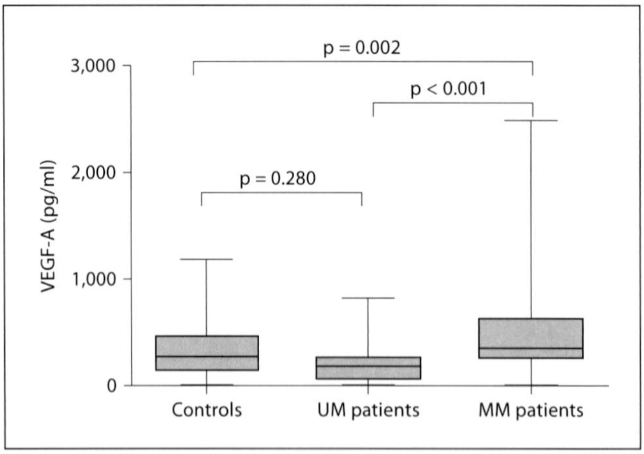

Fig. 2. Concentration of serum VEGF-A in (metastatic) uveal melanoma patients and controls. Concentration of serum VEGF-A in the control group (n = 50), and in patients with (n = 20) and without metastatic (n = 74) uveal melanoma. p values (Mann-Whitney test) between the different groups are indicated in the graph. Each box shows the median, quartiles (box length is the interquartile range) and whiskers represent the 90th and 10th percentiles (from el Filali et al. [9]). UM = Uveal melanoma; MM = metastatic melanoma.

vivo, in a model resembling the tumor environment including paracrine signaling of endothelial cells.

In several tumors (e.g. colon carcinoma, soft tissue sarcomas and gastric cancer), serum VEGF levels have been found to be a marker of disease stage and an indicator of metastases [5, 6, 72]. Until recently, lactate dehydrogenase and alkaline phosphatase were the most indicative serum markers for metastatic disease in uveal melanoma, in combination with liver ultrasonography [73, 74]. Elevated serum osteopontin, melanoma-inhibitory activity and S-100β levels showed a correlation with metastatic uveal melanoma to the liver in some studies [75, 76]. However, serum markers that indicate micrometastases at an early stage would be clinically preferable. In contrast to the immunohistochemical study of Sheidow et al. [69], several studies have observed VEGF expression in melanoma cell lines to be correlated with development of experimental metastasis [77, 78]. In uveal melanoma, we found no difference in the amount of VEGF in sera of uveal melanoma patients compared to healthy people. However, VEGF levels are significantly raised in uveal melanoma patients with metastases compared with those without metastic disease ($p < 0.001$) (fig. 2). The same finding has recently been confirmed in other studies [79]. In addition, using a uveal melanoma mouse model, VEGF serum levels were increased in the presence of hepatic micrometastases in hypoxic regions of the liver [80]. Also, Barak et al. [8] demonstrated a significant increase of VEGF in sera of uveal melanoma patients after the occurrence of metastases; however, wide inter-patient variance prevents the use of a single VEGF serum level to be used as a marker for metastatic disease.

Approved Anti-Angiogenic Treatment in Cancer

In the last two decades, most of the anticancer angiogenic treatments have focused on VEGF/VEGFR and EGF/EGFR, since these factors play such an important role in tumor angiogenesis. There are several other anti-angiogenic drugs that have been approved for the treatment of several different tumors (table 1). The four main methods used to block VEGF or any other angiogenic factors are:

1 neutralizing monoclonal antibodies against the factor or its receptor: bevacizumab, cetuximab, panitumumab, trastuzumab, ranibuzimab;
2 small molecule tyrosine kinase inhibitors (TKIs) of receptors: sorafenib, sunitinib, erlotinib;
3 soluble receptors which act as decoy receptors: VEGF-Trap;
4 ribozymes which specifically target mRNA.

Due to the complexity of angiogenesis, as reviewed in the section on tumor angiogenesis, it is obvious that there may be several indirect ways to inhibit vessel growth besides the direct blocking of angiogenic factors. Temsirolimus (Torisel®), for instance, is an mTOR inhibitor that has direct antitumor activity by arresting cells in the G1 phase of the cell cycle and increasing apoptosis, but that also suppresses HIF-1α transcription levels in tumor cells, thus reducing VEGF expression and angiogenesis [81]. Bortezomib (Velcade®) is a proteasome inhibitor that has been shown to inhibit VEGF, IGF-1, and angiopoietin by an unknown mechanism in multiple myeloma [82]. Thalidomide (Thalomid®) has been unpopular since the 1960s when its severe teratogenic effects were unknown and its use resulted in malformations of the extremities in unborn children of pregnant users. Nevertheless, thalidomide (or its derivative lenalidomide, introduced in 2004) has recently been shown to have potent anti-angiogenic properties, such as decreasing vascular density and successfully blocking angiogenic factors such as basic fibroblast growth factor, and VEGF and it is now under investigation for suppressing tumor angiogenesis [83]. Other approved and established drugs that have been found to exert anti-angiogenic activity include doxycycline [84] and celecoxcib [85].

Additionally, multiple other agents targeting tumor angiogenesis in several different ways are still in (pre-)clinical investigation and should provide more treatment options in the future.

Anti-VEGF antibodies (bevacizumab) and TKIs (sunitinib and sorafinib) will be highlighted in the next section.

Anti-Angiogenic Therapy in Uveal Melanoma

At this time, no anti-angiogenic drugs are used clinically for the treatment of uveal melanoma or its metastases. Intravitreal use of bevacizumab in three cases of uveal melanoma who were wrongfully diagnosed as choroidal neovascularizations did not demonstrate inhibition of tumor growth [86].

Table 1. Angiogenic inhibitors approved for tumor treatment

Drug	Method	Indications	Treatment
Bevacizumab (Avastin®)	Humanized anti-VEGF monoclonal antibody	Colorectal cancer, Non-small cell lung cancer, Advanced breast cancer	In combination with 5-FU-chemotherapy, carboplatin and paclitaxel
Sorafenib (Nexavar®)	Small molecule TK inhibitor of VEGFR-1, VEGFR-2, VEGFR-3, PDGFR-ß, and Raf-1	Advanced renal cell carcinoma, Advanced hepatocellular carcinoma	Monotherapy
Sunitinib (Sutent®)	Small molecule TK inhibitor of VEGFR-1, VEGFR-2, VEGFR-3, PDGFR- ß, and RET	Gastrointestinal stromal tumor, Advanced renal cell carcinoma	Monotherapy
Panitumumab (Vectibix®)	Humanized IgG2 anti-EGFR monoclonal antibody	Metastatic colorectal cancer	Monotherapy after failed chemotherapy with fluoropyrimidine, oxaliplatin, and irinotecan
Cetuximab (Erbitux®)	IgG1 anti-EGFR monoclonal antibody	Metastatic colorectal cancer, Head and neck cancer	Monotherapy and in combination with irinotecan and radiation
Erlotinib (Tarveca®)	Small molecule TK inhibitor of EGFR	Non-small cell lung cancer, Pancreatic cancer	Monotherapy after failed chemotherapy and combination with gemcitabine
Trastuzumab (Herceptin®)	Humanized IgG1 anti- HER-2 monoclonal antibody	Breast cancer	Monotherapy and combination with doxorubicin, cyclophosphamide, and paclitaxel
Temsirolimus (Torisel®)	A small molecule inhibitor of mTOR, supression of HIF-1a	Advanced renal cell carcinoma	Monotherapy
Bortezomib (Velcade®)	Proteasome inhibitor; antiangiogenic (inhibition VEGF, IGF, Ang; mechanism unclear	Multiple myeloma, Mantle cell lymphoma	Monotherapy after failed treatment
Thalidomide (Thalomid®)	Immunomodulatory, antiangiogenic properties, mechanism unclear	Multiple myeloma	Combination with dexamethasone

With regard to the primary tumor, the current treatment includes enucleation, local resection and radiotherapy, either by brachytherapy (iodine or ruthenium), stereotactic or proton beam irradiation [87–90]. Radiotherapy achieves local tumor control in up to 97% of all treated eyes, and can therefore be regarded as being very effective [91, 92]. One could therefore argue whether other therapies, such as anti-angiogenic drugs, are necessary. First of all, not all tumors can be irradiated: contraindications for irradiation are a tumor height of more than 10.0 mm (or 8.0 mm near the disk), a tumor diameter of more than 16.0 mm (although in the case of proton beam irradiation, larger tumors can be treated), when the tumor is not clearly defined by echography, is diffuse or multifocal, when there is neovascular or secondary glaucoma and in case of extrascleral extension [87, 90]. In addition, radiation therapy can lead to radiation retinopathy, a delayed-onset complication characterized by retinal ischemia, neovascularization and leaking vessels [93, 94]. Ultimately, radiation retinopathy can result in a severe decrease of visual acuity in the 'preserved' tumor-containing eye.

Anti-Angiogenic Treatment of Radiation Retinopathy

In eyes with uveal melanoma, bevacizumab is frequently utilized for the treatment of radiation retinopathy. Radiation retinopathy has been described in up to 63% of eyes after plaque radiation [95, 96]. 'Off-label' use of intravitreal bevacizumab to treat macular edema and neovascularization in radiation retinopathy demonstrates a decrease of macular edema and an improvement of visual acuity [97–99].

Other anti-VEGF agents besides bevacizumab have been widely used in ophthalmology this last decade in the treatment of age-related macular degeneration, diabetic macular edema and neovascular glaucoma. Pegaptanib (Macugen), an aptamer that only binds $VEGF_{165}$, was the first drug to receive approval for the treatment of macular degeneration. Although this drug is hardly used for any ocular pathology, one case study describes improved visual acuity after treatment with pegaptanib in a patient with proliferative radiation retinopathy [100]. Ranibizumab (Lucentis) is a recombinant humanized immunoglobulin monoclonal antibody fragment especially designed for intraocular use and is approved in many countries. The efficacy and safety of ranibizumab were evaluated in several randomized trials involving more than 1,000 patients with neovascular age-related macular degeneration and was shown to significantly maintain (90%) and improve (33%) visual acuity after 24 months [101]. In addition, treatment with ranibizumab also improved visual acuity in 4 of 5 patients with radiation maculopathy [102]. VEGF-Trap is a soluble protein that acts as a VEGF decoy receptor, and is currently undergoing phase 3 testing for age-related macular degeneration as well as for metastatic melanoma treatment [103].

Anti-Angiogenic Therapy for Uveal Melanoma Metastasis

Almost 50% of all uveal melanoma patients eventually develop metastatic disease, with the current 5-year uveal melanoma-related mortality ranging from 26 to 32% [27]. Life expectancy in the case of uveal melanoma-related metastatic disease ranges from 2 to 6 months since hardly any effective treatment is currently available; chemotherapy and local resection only prolong survival by a few months [104]. A number of anti-angiogenic agents may be of clinical use.

Bevacizumab

Yang et al. [16] studied the effect of bevacizumab on the growth inhibition and number of hepatic micrometastases in an ocular melanoma mouse model, in which B16 melanoma cells were inoculated subchoroidally. Bevacizumab was administered by intraperitoneal injection (starting dose: 50 or 250 μg/100 μl). Bevacizumab suppressed primary ocular melanoma growth and the formation of hepatic micrometastases in a dose-dependent manner ($p < 0.01$). In addition, bevacizumab significantly reduced the level of VEGF in the culture media of two human uveal melanoma cell lines.

In contrast, we found a rather unexpected effect of bevacizumab on uveal melanoma. Our mouse model consisted of B16 melanoma cells which were placed into the anterior chamber of the eye and bevacizumab (10 times the equivalent human dose, 20 μg/4 μl; equivalent human dose, 2 μg/4 μl) or a mock injection were given intraocularly. In vivo acceleration of intraocular tumor growth was observed in the eyes treated with bevacizumab, although it did not influence B16 or uveal melanoma cell proliferation in vitro. Remarkably, bevacizumab did increase mRNA VEGF melanoma expression and HIF-1α stabilization in vitro. This was especially seen under hypoxic conditions. Only after treatment with bevacizumab did we observe anterior chamber and tumor hemorrhages in murine eyes, emphasizing increased microvascular permeability, possibly due to induced VEGF expression. This 'pseudohypoxic' phenomenon has been described in other tumors and may be the consequence of a tumor adaptive or evasive response. It will be further elaborated in the following section [el Filali et al., submitted].

Sorafenib

Sorafenib, which inhibits VEGFR, has been tested in a xenograft model in which uveal melanoma cell line 92.1 was injected subcutaneously. Mangiameli et al. [105] demonstrated inhibition of tumor growth ($p < 0.0035$) and fewer metastases after sorafenib treatment (33 vs. 60%). In patients with metastatic cutaneous melanoma, monotherapy with sorafenib has demonstrated hardly any antitumor activity [106]. Recently, the final results of a phase 3 trial, which compared treatment of metastatic (not including uveal) melanoma patients ($n = 823$) with carboplatin, paclitaxel and with either sorafenib (SCP) or a placebo (CP), did not demonstrate a difference in overall survival: the median overall survival for the SCP group was 11.1 months (95% CI 10.3–12.3) and for the CP group 11.3 months (95% CI 9.7–12.3) (ASCO meeting 2010, abstract number 8511).

Sunitinib

Sunitinib is another TKI which inhibits VEGFR [107]. There is not much preclinical evidence for antitumor activity in uveal melanoma. Still, a clinical benefit in advanced metastatic melanoma patients has recently been observed in a phase 2 trial analyzing the effect of sunitinib monotherapy. Three patients (8.3%) demonstrated a partial response, with a mean duration of 6.5 months. Nine had stable disease (25%), with a mean duration of 4.1 months (range: 3–8.2 months), and 17 had progressive disease (47.2%) (ASCO meeting 2010, abstract number 8518).

Although uveal melanoma is the most common intraocular tumor in adults (0.7/100,000 per year), it is still a relatively rare form of cancer. Conducting a good clinical trial in such a population is therefore quite challenging. There are ongoing and recruiting trials investigating bevacizumab, sorafenib and sunitinib as a single agent or in combination with other regimes. Most studies are focused on cutaneous metastatic melanoma. Fortunately, trials currently also include uveal melanoma patients and some of them even enroll only patients with ocular melanoma-related metastasis (clinicaltrials.gov). Hopefully this will give us some insight into current clinical anti-angiogenic treatments.

Adverse Effects of Anti-Angiogenic Therapy

Vascular Endothelial Growth Factor Inhibitors

Clinical experience with predominantly bevacizumab has revealed that anti-VEGF therapy often prolongs overall survival of cancer patients by a few months, without really curing metastatic disease [108]. It has been proposed these past few years that VEGF inhibitors may actually promote tumorigenesis and metastatic dissemination on the long run [15, 109]. Recently, the FDA has prohibited the use of bevacizumab (monotherapy) for metastatic breast cancer patients. When we observed tumor acceleration after treatment with bevacizumab in our mouse model, we also analyzed the effect on human uveal melanoma cells. In vitro treatment with bevacizumab induced VEGF mRNA expression in uveal melanoma cells. Moreover, we observed that this upregulation involved the HIF-1α pathway [el Filali et al., submitted]. VEGF inhibitors seem to elicit similar effects as described earlier for ischemic conditions that induce VEGF expression in uveal melanoma cells through the HIF-1α pathway. The paradox of VEGF upregulation upon anti-VEGF treatment has been dubbed 'pseudohypoxia' and has been described before in other tumor studies. In mice bearing intracerebral glioma, it has been demonstrated that anti-VEGF treatment with pegaptanib (Macugen) increases GLUT-1 expression (glucose transporter also upregulated by HIF-1α) [51]. This 'pseudohypoxia' has also been shown to increase tumorigenesis in other types of cancer cells. Treatment of mice with pancreatic neuroendocrine tumors with anti-VEGFR also resulted in an initial response of tumor stasis followed by tumor recurrence. The relapsing tumor expressed higher levels of mRNAs of proangiogenic factors and demonstrated several hypoxic regions [110]. Furthermore, treated mice developed more invasive tumors and metastatic lesions, all characterized by hypoxic regions [15].

Ischemic conditions caused by anti-VEGF treatment can also lead to recruitment of various bone marrow-derived cells that have angiogenic capacities. Proangiogenic monocytes induce vessel growth by expression of several cytokines and angiogenic factors. In mice bearing glioblastoma multiforme tumors treated with bevacizumab, the stabilization of HIF-1α has been demonstrated to promote angiogenesis by inducing recruitment of mature F4/80+ macrophages [111, 112]. Additionally, a clinical study suggests that hypoxia determines survival outcome in patients treated with bevacizumab for glioblastoma multiforme [113]. Since it has previously been shown that malignant uveal melanoma tumors in patients with a poor survival contain many macrophages, this mechanism is especially relevant [114].

Moreover, we may be observing in our experiments resistance of the tumor cells, after an initial response phase, to adapt or evade therapy by inducing mechanisms that reduce dependence on neovascularization, leading to changed tumor proliferation. The 'pseudohypoxic' conditions could be responsible for selection of more malignant tumor cells, which are less sensitive to anti-angiogenic treatment and switch on other malignant pathways that result in proliferation, migration and invasion. Besides angiogenesis and vascular permeability, VEGF has been shown to activate the ras/raf pathway through activation of the tyrosine kinase VEGF receptors and the downstream MAPKs [115, 116]. MAPK-driven proliferation has been shown to play an important role in uveal melanoma growth through upstream Src signaling [59].

Tyrosine Kinase Inhibitors

TKI side effects are related to their nonspecific nature [117]. In order to be able to predict treatment outcome, one should know the effect of TKIs on the different pathways and how these pathways may interact. Sorafenib treatment of cutaneous melanoma patients may have been disappointing because the combined effect on all inhibited kinases turned out to be negative for tumor inhibition [118]. Although the capacity of TKIs to target multiple kinases is interesting because of its wide application in several different malignancies, it also results in many 'off-target' effects demonstrated in several clinical trials. For example, hand-foot skin reactions, fatigue, stomatitis, diarrhea, hair color changes, myelosuppression, and thyroid dysfunction are frequently associated with TKI treatment. In addition, the low effectiveness of available TKIs requires higher doses. Unfortunately, higher doses are in turn associated with increased blockade of nontarget kinases due to low selectivity, again resulting in toxicity. The off-target effects of TKIs have also limited their use in combination with chemotherapeutic drugs due to overlapping toxicity profiles. Recently, treatment with sunitinib and sorafenib has been associated with cardiovascular toxicity as an adverse event [119]. These limitations have led to the development of more selective and potent anti-VEGFR TKIs [120].

Conclusion

Uveal melanoma remains a highly lethal tumor, which results in metastatic disease in almost 50% of all patients despite adequate primary tumor treatment. It is therefore important to investigate different treatment options to be used in curative or preventive therapy of uveal melanoma-related metastatic disease. Tumor angiogenesis has been demonstrated to be of great importance in tumor growth and dissemination. In addition, tumor vessel formation is also complex and extensive, involving various molecular mediators and pathways. Anti-angiogenic therapy has focused on VEGF, which has been implicated in uveal melanoma angiogenesis. Unfortunately, experimental and clinical trials using anti-VEGF monotherapy have been disappointing. In addition, VEGF inhibitors may actually promote tumorigenesis and metastatic dissemination. The key to effective treatment is good patient and tumor selection. The inhibition of protein activity by small molecules appears to be a promising approach for several types of malignancies. For example, imatinib has been analyzed for treatment of uveal melanoma-related metastases in a clinical trial, based on c-Kit overexpression and the in vitro response of cell lines with c-Kit expression to imatinib mesylate [121]. Treatment with imatinib mesylate did not result in improved survival, which may be due to absence of c-Kit upregulation in the patients in the trial because patients had been treated irrespective of their c-Kit tumor status [122, 123]. This could also be the case with anti-angiogenic therapies in which patients are treated irrespective of the angiogenic profile and VEGF/VEGFR expression of their uveal melanoma and/or metastases. We demonstrated that angiogenesis and especially VEGF expression can easily be modulated by the uveal melanoma cells themselves, either by tumor microenvironment or due to VEGF inhibitors. In addition, structures identified as vasculogenic mimicry may provide uveal melanoma with an alternative tumor circulation. Therefore, one could still question whether tumor angiogenesis and the angiogenic switch are necessary for uveal melanoma growth and malignant dissemination. They may merely be a consequence of tumor growth.

References

1 Holmgren L, O'Reilly MS, Folkman J: Dormancy of micrometastases: balanced proliferation and apoptosis in the presence of angiogenesis suppression. Nat Med 1995;1:149–153.

2 Papetti M, Herman IM: Mechanisms of normal and tumor-derived angiogenesis. Am J Physiol Cell Physiol 2002;282:C947–C970.

3 Criscuolo GR, Merrill MJ, Oldfield EH: Further characterization of malignant glioma-derived vascular permeability factor. J Neurosurg 1988;69:254–262.

4 Ferrara N, Gerber HP, LeCouter J: The biology of VEGF and its receptors. Nat Med 2003;9:669–676.

5 Karayiannakis AJ, Syrigos KN, Polychronidis A, Zbar A, Kouraklis G, Simopoulos C, Karatzas G: Circulating VEGF levels in the serum of gastric cancer patients: correlation with pathological variables, patient survival, and tumor surgery. Ann Surg 2002;236:37–42.

6 Celen O, Kahraman I, Yildirim E, Berberoglu U: Correlation of vascular endothelial growth factor (VEGF) and CEA with clinicopathological variables in colorectal cancer patients. Neoplasma 2004;51:293–299.

7 Hayes AJ, Mostyn-Jones A, Koban MU, A'Hern R, Burton P, Thomas JM: Serum vascular endothelial growth factor as a tumour marker in soft tissue sarcoma. Br J Surg 2004;91:242–247.

8 Barak V, Pe'er J, Kalickman I, Frenkel S: VEGF as a biomarker for metastatic uveal melanoma in humans. Curr Eye Res 2011;36:386–390.

9 el Filali M, Missotten GS, Maat W, Ly LV, Luyten GP, van der Velden PA, Jager MJ: Regulation of VEGF-A in uveal melanoma. Invest Ophthalmol Vis Sci 2010;51:2329–2337.

10 Kabbinavar F, Hurwitz HI, Fehrenbacher L, Meropol NJ, Novotny WF, Lieberman G, Griffing S, Bergsland E: Phase II, randomized trial comparing bevacizumab plus fluorouracil (FU)/leucovorin (LV) with FU/LV alone in patients with metastatic colorectal cancer. J Clin Oncol 2003;21:60–65.

11 Yang JC, Haworth L, Sherry RM, Hwu P, Schwartzentruber DJ, Topalian SL, Steinberg SM, Chen HX, Rosenberg SA: A randomized trial of bevacizumab, an anti-vascular endothelial growth factor antibody, for metastatic renal cancer. N Engl J Med 2003;349:427–434.

12 Cobleigh MA, Langmuir VK, Sledge GW, Miller KD, Haney L, Novotny WF, Reimann JD, Vassel A: A phase I/II dose-escalation trial of bevacizumab in previously treated metastatic breast cancer. Semin Oncol 2003;30:117–124.

13 Varker KA, Biber JE, Kefauver C, Jensen R, Lehman A, Young D, Wu H, Lesinski GB, Kendra K, Chen HX, Walker MJ, Carson WE 3rd: A randomized phase 2 trial of bevacizumab with or without daily low-dose interferon alfa-2b in metastatic malignant melanoma. Ann Surg Oncol 2007;14:2367–2376.

14 Van CE, Vervenne WL, Bennouna J, Humblet Y, Gill S, Van Laethem JL, Verslype C, Scheithauer W, Shang A, Cosaert J, Moore MJ: Phase III trial of bevacizumab in combination with gemcitabine and erlotinib in patients with metastatic pancreatic cancer. J Clin Oncol 2009;27:2231–2237.

15 Paez-Ribes M, Allen E, Hudock J, Takeda T, Okuyama H, Vinals F, Inoue M, Bergers G, Hanahan D, Casanovas O: Antiangiogenic therapy elicits malignant progression of tumors to increased local invasion and distant metastasis. Cancer Cell 2009; 15:220–231.

16 Yang H, Jager MJ, Grossniklaus HE: Bevacizumab suppression of establishment of micrometastases in experimental ocular melanoma. Invest Ophthalmol Vis Sci 2010;51:2835–2842.

17 Foss AJ, Alexander RA, Jefferies LW, Hungerford JL, Harris AL, Lightman S: Microvessel count predicts survival in uveal melanoma. Cancer Res 1996;56: 2900–2903.

18 Lane AM, Egan KM, Yang J, Saornil MA, Alroy J, Albert D, Gragoudas ES: An evaluation of tumour vascularity as a prognostic indicator in uveal melanoma. Melanoma Res 1997;7:237–242.

19 Foss AJ, Alexander RA, Hungerford JL, Harris AL, Cree IA, Lightman S: Reassessment of the PAS patterns in uveal melanoma. Br J Ophthalmol 1997;81: 240–246.

20 Makitie T, Summanen P, Tarkkanen A, Kivela T: Microvascular density in predicting survival of patients with choroidal and ciliary body melanoma. Invest Ophthalmol Vis Sci 1999;40:2471–2480.

21 Makitie T, Summanen P, Tarkkanen A, Kivela T: Microvascular loops and networks as prognostic indicators in choroidal and ciliary body melanomas. J Natl Cancer Inst 1999;91:359–367.

22 Folberg R, Hendrix MJ, Maniotis AJ: Vasculogenic mimicry and tumor angiogenesis. Am J Pathol 2000; 156:361–381.

23 Shirakawa K, Tsuda H, Heike Y, Kato K, Asada R, Inomata M, Sasaki H, Kasumi F, Yoshimoto M, Iwanaga T, Konishi F, Terada M, Wakasugi H: Absence of endothelial cells, central necrosis, and fibrosis are associated with aggressive inflammatory breast cancer. Cancer Res 2001;61:445–451.

24 Coupland SE, Campbell I, Damato B: Routes of extraocular extension of uveal melanoma: risk factors and influence on survival probability. Ophthalmology 2008;115:1778–1785.

25 Ly LV, Odish OF, Wolff-Rouendaal D, Missotten GS, Luyten GP, Jager MJ: Intravascular presence of tumor cells as prognostic parameter in uveal melanoma: a 35-year survey. Invest Ophthalmol Vis Sci 2010;51:658–665.

26 Augsburger JJ, Correa ZM, Freire J, Brady LW: Long-term survival in choroidal and ciliary body melanoma after enucleation versus plaque radiation therapy. Ophthalmology 1998;105:1670–1678.

27 Kujala E, Makitie T, Kivela T: Very long-term prognosis of patients with malignant uveal melanoma. Invest Ophthalmol Vis Sci 2003;44:4651–4659.

28 Economou MA, All-Ericsson C, Bykov V, Girnita L, Bartolazzi A, Larsson O, Seregard S: Receptors for the liver synthesized growth factors IGF-1 and HGF/SF in uveal melanoma: intercorrelation and prognostic implications. Invest Ophthalmol Vis Sci 2005;46:4372–4375.

29 Baserga R: The insulin-like growth factor I receptor: a key to tumor growth? Cancer Res 1995;55:249–252.

30 Slomiany MG, Rosenzweig SA: IGF-1-induced VEGF and IGFBP-3 secretion correlates with increased HIF-1 alpha expression and activity in retinal pigment epithelial cell line D407. Invest Ophthalmol Vis Sci 2004;45:2838–2847.

31 Bergers G, Benjamin LE: Tumorigenesis and the angiogenic switch. Nat Rev Cancer 2003;3:401–410.

32 Handsley MM, Edwards DR: Metalloproteinases and their inhibitors in tumor angiogenesis. Int J Cancer 2005;115:849–860.

33 Littlepage LE, Sternlicht MD, Rougier N, Phillips J, Gallo E, Yu Y, Williams K, Brenot A, Gordon JI, Werb Z: Matrix metalloproteinases contribute distinct roles in neuroendocrine prostate carcinogenesis, metastasis, and angiogenesis progression. Cancer Res 2010;70:2224–2234.

34 El Shabrawi Y, Ardjomand N, Radner H, Ardjomand N: MMP-9 is predominantly expressed in epithelioid and not spindle cell uveal melanoma. J Pathol 2001;194:201–206.
35 De Luca A, Carotenuto A, Rachiglio A, Gallo M, Maiello MR, Aldinucci D, Pinto A, Normanno N: The role of the EGFR signaling in tumor microenvironment. J Cell Physiol 2008;214:559–567.
36 Ferrara N: The role of VEGF in the regulation of physiological and pathological angiogenesis. EXS 2005;209–231.
37 Boyd SR, Tan D, Bunce C, Gittos A, Neale MH, Hungerford JL, Charnock-Jones S, Cree IA: Vascular endothelial growth factor is elevated in ocular fluids of eyes harbouring uveal melanoma: identification of a potential therapeutic window. Br J Ophthalmol 2002;86:448–452.
38 Ferrara N, Davis-Smyth T: The biology of vascular endothelial growth factor. Endocr Rev 1997;18:4–25.
39 Houck KA, Ferrara N, Winer J, Cachianes G, Li B, Leung DW: The vascular endothelial growth factor family: identification of a fourth molecular species and characterization of alternative splicing of RNA. Mol Endocrinol 1991;5:1806–1814.
40 Keyt BA, Nguyen HV, Berleau LT, Duarte CM, Park J, Chen H, Ferrara N: Identification of vascular endothelial growth factor determinants for binding KDR and FLT-1 receptors. Generation of receptor-selective VEGF variants by site-directed mutagenesis. J Biol Chem 1996;271:5638–5646.
41 Weigand M, Hantel P, Kreienberg R, Waltenberger J: Autocrine vascular endothelial growth factor signalling in breast cancer. Evidence from cell lines and primary breast cancer cultures in vitro. Angiogenesis 2005;8:197–204.
42 Ferrara N: Vascular endothelial growth factor as a target for anticancer therapy. Oncologist 2004; 9(suppl 1):2–10.
43 Vincenti V, Cassano C, Rocchi M, Persico G: Assignment of the vascular endothelial growth factor gene to human chromosome 6p21.3. Circulation 1996;93:1493–1495.
44 Prescher G, Bornfeld N, Becher R: Nonrandom chromosomal abnormalities in primary uveal melanoma. J Natl Cancer Inst 1990;82:1765–1769.
45 Naus NC, van Drunen E, de Klein A, Luyten GP, Paridaens DA, Alers JC, Ksander BR, Beverloo HB, Slater RM: Characterization of complex chromosomal abnormalities in uveal melanoma by fluorescence in situ hybridization, spectral karyotyping, and comparative genomic hybridization. Genes Chromosomes Cancer 2001;30:267–273.
46 Abdel-Rahman MH, Craig EL, Davidorf FH, Eng C: Expression of vascular endothelial growth factor in uveal melanoma is independent of 6p21-region copy number. Clin Cancer Res 2005;11:73–78.
47 White VA, Chambers JD, Courtright PD, Chang WY, Horsman DE: Correlation of cytogenetic abnormalities with the outcome of patients with uveal melanoma. Cancer 1998;83:354–359.
48 Shweiki D, Itin A, Soffer D, Keshet E: Vascular endothelial growth factor induced by hypoxia may mediate hypoxia-initiated angiogenesis. Nature 1992;359:843–845.
49 Wang GL, Semenza GL: Purification and characterization of hypoxia-inducible factor 1. J Biol Chem 1995;270:1230–1237.
50 Mizukami Y, Kohgo Y, Chung DC: Hypoxia inducible factor-1 independent pathways in tumor angiogenesis. Clin Cancer Res 2007;13:5670–5674.
51 Verhoeff JJ, Stalpers LJ, Claes A, Hovinga KE, Musters GD, Peter VW, Richel DJ, Leenders WP, van Furth WR: Tumour control by whole brain irradiation of anti-VEGF-treated mice bearing intracerebral glioma. Eur J Cancer 2009;45:3074–3080.
52 Shibuya M: Structure and function of VEGF/VEGF-receptor system involved in angiogenesis. Cell Struct Funct 2001;26:25–35.
53 Economou MA, Wu J, Vasilcanu D, Rosengren L, All-Ericsson C, van der Ploeg I, Menu E, Girnita L, Axelson M, Larsson O, Seregard S, Kvanta A: Inhibition of VEGF secretion and experimental choroidal neovascularization by picropodophyllin (PPP), an inhibitor of the insulin-like growth factor-1 receptor. Invest Ophthalmol Vis Sci 2008;49:2620–2626.
54 Economou MA, Andersson S, Vasilcanu D, All-Ericsson C, Menu E, Girnita A, Girnita L, Axelson M, Seregard S, Larsson O: Oral picropodophyllin (PPP) is well tolerated in vivo and inhibits IGF-1R expression and growth of uveal melanoma. Invest Ophthalmol Vis Sci 2008;49:2337–2342.
55 Siemeister G, Weindel K, Mohrs K, Barleon B, Martiny-Baron G, Marme D: Reversion of deregulated expression of vascular endothelial growth factor in human renal carcinoma cells by von Hippel-Lindau tumor suppressor protein. Cancer Res 1996;56:2299–2301.
56 Rankin EB, Rha J, Unger TL, Wu CH, Shutt HP, Johnson RS, Simon MC, Keith B, Haase VH: Hypoxia-inducible factor-2 regulates vascular tumorigenesis in mice. Oncogene 2008;27:5354–5358.
57 Haase VH, Glickman JN, Socolovsky M, Jaenisch R: Vascular tumors in livers with targeted inactivation of the von Hippel-Lindau tumor suppressor. Proc Natl Acad Sci USA 2001;98:1583–1588.

58 Rak J, Mitsuhashi Y, Bayko L, Filmus J, Shirasawa S, Sasazuki T, Kerbel RS: Mutant ras oncogenes upregulate VEGF/VPF expression: implications for induction and inhibition of tumor angiogenesis. Cancer Res 1995;55:4575–4580.
59 Maat W, el Filali M, Dirsk-Mulder A, Luyten GP, Gruis NA, Desjardins L, Boender P, Jager MJ, van der Velden PA: Episodic Src activation in uveal melanoma revealed by kinase activity profiling. Br J Cancer 2009;101:312–319.
60 Takahashi H, Shibuya M: The vascular endothelial growth factor (VEGF)/VEGF receptor system and its role under physiological and pathological conditions. Clin Sci (Lond) 2005;109:227–241.
61 Bates DO, Harper SJ: Regulation of vascular permeability by vascular endothelial growth factors. Vascul Pharmacol 2002;39:225–237.
62 Qi JH, Matsumoto T, Huang K, Olausson K, Christofferson R, Claesson-Welsh L: Phosphoinositide 3 kinase is critical for survival, mitogenesis and migration but not for differentiation of endothelial cells. Angiogenesis 1999;3:371–380.
63 Jager MJ, Ly LV, El Filali M, Madigan MC: Macrophages in uveal melanoma and in experimental ocular tumor models: friends or foes? Prog Retin Eye Res 2011;30:129–146.
64 Volm M, Koomagi R, Mattern J, Stammler G: Angiogenic growth factors and their receptors in non-small cell lung carcinomas and their relationships to drug response in vitro. Anticancer Res 1997;17:99–103.
65 Brown LF, Berse B, Jackman RW, Tognazzi K, Manseau EJ, Senger DR, Dvorak HF: Expression of vascular permeability factor (vascular endothelial growth factor) and its receptors in adenocarcinomas of the gastrointestinal tract. Cancer Res 1993;53:4727–4735.
66 Yoshiji H, Gomez DE, Shibuya M, Thorgeirsson UP: Expression of vascular endothelial growth factor, its receptor, and other angiogenic factors in human breast cancer. Cancer Res 1996;56:2013–2016.
67 Witmer AN, Vrensen GF, Van Noorden CJ, Schlingemann RO: Vascular endothelial growth factors and angiogenesis in eye disease. Prog Retin Eye Res 2003;22:1–29.
68 Ijland SA, Jager MJ, Heijdra BM, Westphal JR, Peek R: Expression of angiogenic and immunosuppressive factors by uveal melanoma cell lines. Melanoma Res 1999;9:445–450.
69 Sheidow TG, Hooper PL, Crukley C, Young J, Heathcote JG: Expression of vascular endothelial growth factor in uveal melanoma and its correlation with metastasis. Br J Ophthalmol 2000;84:750–756.
70 Boyd SR, Tan DS, de Souza L, Neale MH, Myatt NE, Alexander RA, Robb M, Hungerford JL, Cree IA: Uveal melanomas express vascular endothelial growth factor and basic fibroblast growth factor and support endothelial cell growth. Br J Ophthalmol 2002;86:440–447.
71 Missotten GS, Notting IC, Schlingemann RO, Zijlmans HJ, Lau C, Eilers PH, Keunen JE, Jager MJ: Vascular endothelial growth factor a in eyes with uveal melanoma. Arch Ophthalmol 2006;124:1428–1434.
72 Hayes AJ, Mostyn-Jones A, Koban MU, A'Hern R, Burton P, Thomas JM: Serum vascular endothelial growth factor as a tumour marker in soft tissue sarcoma. Br J Surg 2004;91:242–247.
73 Eskelin S, Pyrhonen S, Summanen P, Prause JU, Kivela T: Screening for metastatic malignant melanoma of the uvea revisited. Cancer 1999;85:1151–1159.
74 Kaiserman I, Amer R, Pe'er J: Liver function tests in metastatic uveal melanoma. Am J Ophthalmol 2004; 137:236–243.
75 Barak V, Frenkel S, Kalickman I, Maniotis AJ, Folberg R, Pe'er J: Serum markers to detect metastatic uveal melanoma. Anticancer Res 2007;27:1897–1900.
76 Missotten GS, Korse CM, van Dehn C, Linders TC, Keunen JE, Jager MJ, Bonfrer JM: S-100B protein and melanoma inhibitory activity protein in uveal melanoma screening. A comparison with liver function tests. Tumour Biol 2007;28:63–69.
77 Potgens AJ, Lubsen NH, van Altena MC, Schoenmakers JG, Ruiter DJ, de Waal RM: Vascular permeability factor expression influences tumor angiogenesis in human melanoma lines xenografted to nude mice. Am J Pathol 1995;146:197–209.
78 Claffey KP, Brown LF, del Aguila LF, Tognazzi K, Yeo KT, Manseau EJ, Dvorak HF: Expression of vascular permeability factor/vascular endothelial growth factor by melanoma cells increases tumor growth, angiogenesis, and experimental metastasis. Cancer Res 1996;56:172–181.
79 Sahin A, Kiratli H, Soylemezoglu F, Tezel GG, Bilgic S, Saracbasi O: Expression of vascular endothelial growth factor-A, matrix metalloproteinase-9, and extravascular matrix patterns and their correlations with clinicopathologic parameters in posterior uveal melanomas. Jpn J Ophthalmol 2007;51:325–331.
80 Crosby MB, Yang H, Gao W, Zhang L, Grossniklaus HE: Serum vascular endothelial growth factor (VEGF) levels correlate with number and location of micrometastases in a murine model of uveal melanoma. Br J Ophthalmol 2011;95:112–117.

81 Land SC, Tee AR: Hypoxia-inducible factor 1alpha is regulated by the mammalian target of rapamycin (mTOR) via an mTOR signaling motif. J Biol Chem 2007;282:20534–20543.

82 Roccaro AM, Hideshima T, Raje N, Kumar S, Ishitsuka K, Yasui H, Shiraishi N, Ribatti D, Nico B, Vacca A, Dammacco F, Richardson PG, Anderson KC: Bortezomib mediates antiangiogenesis in multiple myeloma via direct and indirect effects on endothelial cells. Cancer Res 2006;66:184–191.

83 Kenyon BM, Browne F, D'Amato RJ: Effects of thalidomide and related metabolites in a mouse corneal model of neovascularization. Exp Eye Res 1997;64:971–978.

84 Kalas W, Gilpin S, Yu JL, May L, Krchnakova H, Bornstein P, Rak J: Restoration of thrombospondin 1 expression in tumor cells harbouring mutant ras oncogene by treatment with low doses of doxycycline. Biochem Biophys Res Commun 2003;310:109–114.

85 Masferrer JL, Leahy KM, Koki AT, Zweifel BS, Settle SL, Woerner BM, Edwards DA, Flickinger AG, Moore RJ, Seibert K: Antiangiogenic and antitumor activities of cyclooxygenase-2 inhibitors. Cancer Res 2000;60:1306–1311.

86 Lima BR, Schoenfield LR, Singh AD: The impact of intravitreal bevacizumab therapy on choroidal melanoma. Am J Ophthalmol 2011;151:323–328.

87 Melia BM, Abramson DH, Albert DM, Boldt HC, Earle JD, Hanson WF, Montague P, Moy CS, Schachat AP, Simpson ER, Straatsma BR, Vine AK, Weingeist TA: Collaborative ocular melanoma study (COMS) randomized trial of I-125 brachytherapy for medium choroidal melanoma. 1. Visual acuity after 3 years COMS report No 16. Ophthalmology 2001;108:348–366.

88 Shields CL, Shields JA, Cater J, Gunduz K, Miyamoto C, Micaily B, Brady LW: Plaque radiotherapy for uveal melanoma: long-term visual outcome in 1,106 consecutive patients. Arch Ophthalmol 2000;118: 1219–1228.

89 ener-West M, Earle JD, Fine SL, Hawkins BS, Moy CS, Reynolds SM, Schachat AP, Straatsma BR: The COMS randomized trial of iodine 125 brachytherapy for choroidal melanoma. 3. Initial mortality findings. COMS Report No 18. Arch Ophthalmol 2001;119:969–982.

90 Gragoudas ES, Lane AM, Munzenrider J, Egan KM, Li W: Long-term risk of local failure after proton therapy for choroidal/ciliary body melanoma. Trans Am Ophthalmol Soc 2002;100:43–48.

91 Shields CL, Cater J, Shields JA, Chao A, Krema H, Materin M, Brady LW: Combined plaque radiotherapy and transpupillary thermotherapy for choroidal melanoma: tumor control and treatment complications in 270 consecutive patients. Arch Ophthalmol 2002;120:933–940.

92 Jampol LM, Moy CS, Murray TG, Reynolds SM, Albert DM, Schachat AP, Diddie KR, Engstrom RE Jr, Finger PT, Hovland KR, Joffe L, Olsen KR, Wells CG: The COMS randomized trial of iodine 125 brachytherapy for choroidal melanoma. 4. Local treatment failure and enucleation in the first 5 years after brachytherapy. COMS report No 19. Ophthalmology 2002;109:2197–2206.

93 Archer DB, Gardiner TA: Ionizing radiation and the retina. Curr Opin Ophthalmol 1994;5:59–65.

94 Guyer DR, Mukai S, Egan KM, Seddon JM, Walsh SM, Gragoudas ES: Radiation maculopathy after proton beam irradiation for choroidal melanoma. Ophthalmology 1992;99:1278–1285.

95 Sia S, Harper C, McAllister I, Perry A: Iodine-I25 episcleral plaque therapy in uveal melanoma. Clin Experiment Ophthalmol 2000;28:409–413.

96 Jensen AW, Petersen IA, Kline RW, Stafford SL, Schomberg PJ, Robertson DM: Radiation complications and tumor control after ^{125}I plaque brachytherapy for ocular melanoma. Int J Radiat Oncol Biol Phys 2005;63:101–108.

97 Mason JO III, Albert MA Jr, Persaud TO, Vail RS: Intravitreal bevacizumab treatment for radiation macular edema after plaque radiotherapy for choroidal melanoma. Retina 2007;27:903–907.

98 Finger PT: Radiation retinopathy is treatable with anti-vascular endothelial growth factor bevacizumab (Avastin). Int J Radiat Oncol Biol Phys 2008;70:974–977.

99 Gupta A, Muecke JS: Treatment of radiation maculopathy with intravitreal injection of bevacizumab (Avastin). Retina 2008;28:964–968.

100 Querques G, Prascina F, Iaculli C, Delle NN: Intravitreal pegaptanib sodium (Macugen) for radiation retinopathy following episcleral plaque radiotherapy. Acta Ophthalmol 2008;86:700–701.

101 Narayanan R, Kuppermann BD, Jones C, Kirkpatrick P: Ranibizumab. Nat Rev Drug Discov 2006;5:815–816.

102 Finger PT, Chin KJ: Intravitreous ranibizumab (lucentis) for radiation maculopathy. Arch Ophthalmol 2010;128:249–252.

103 Fukasawa M, Korc M: Vascular endothelial growth factor-trap suppresses tumorigenicity of multiple pancreatic cancer cell lines. Clin Cancer Res 2004; 10:3327–3332.

104 Augsburger JJ, Correa ZM, Shaikh AH: Effectiveness of treatments for metastatic uveal melanoma. Am J Ophthalmol 2009;148:119–127.

105 Mangiameli DP, Blansfield JA, Kachala S, Lorang D, Schafer PH, Muller GW, Stirling DI, Libutti SK: Combination therapy targeting the tumor microenvironment is effective in a model of human ocular melanoma. J Transl Med 2007;5:38.

106 Eisen T, Ahmad T, Flaherty KT, Gore M, Kaye S, Marais R, Gibbens I, Hackett S, James M, Schuchter LM, Nathanson KL, Xia C, Simantov R, Schwartz B, Poulin-Costello M, O'Dwyer PJ, Ratain MJ: Sorafenib in advanced melanoma: a phase II randomised discontinuation trial analysis. Br J Cancer 2006;95:581–586.

107 Chow LQ, Eckhardt SG: Sunitinib: from rational design to clinical efficacy. J Clin Oncol 2007;25: 884–896.

108 Kerbel RS: Tumor angiogenesis. N Engl J Med 2008;358:2039–2049.

109 Ebos JM, Lee CR, Bogdanovic E, Alami J, Van Slyke P, Francia G, Xu P, Mutsaers AJ, Dumont DJ, Kerbel RS: Vascular endothelial growth factor-mediated decrease in plasma soluble vascular endothelial growth factor receptor-2 levels as a surrogate biomarker for tumor growth. Cancer Res 2008;68: 521–529.

110 Casanovas O, Hicklin DJ, Bergers G, Hanahan D: Drug resistance by evasion of antiangiogenic targeting of VEGF signaling in late-stage pancreatic islet tumors. Cancer Cell 2005;8:299–309.

111 Aghi M, Cohen KS, Klein RJ, Scadden DT, Chiocca EA: Tumor stromal-derived factor-1 recruits vascular progenitors to mitotic neovasculature, where microenvironment influences their differentiated phenotypes. Cancer Res 2006;66:9054–9064.

112 Du R, Lu KV, Petritsch C, Liu P, Ganss R, Passegue E, Song H, Vandenberg S, Johnson RS, Werb Z, Bergers G: HIF1alpha induces the recruitment of bone marrow-derived vascular modulatory cells to regulate tumor angiogenesis and invasion. Cancer Cell 2008;13:206–220.

113 Sathornsumetee S, Cao Y, Marcello JE, Herndon JE, McLendon RE, Desjardins A, Friedman HS, Dewhirst MW, Vredenburgh JJ, Rich JN: Tumor angiogenic and hypoxic profiles predict radiographic response and survival in malignant astrocytoma patients treated with bevacizumab and irinotecan. J Clin Oncol 2008;26:271–278.

114 Bronkhorst IH, Ly LV, Jordanova ES, Vrolijk H, Versluis M, Luyten GP, Jager MJ: Detection of M2 macrophages in uveal melanoma and relation with survival. Invest Ophthalmol Vis Sci 2011;52:643–650.

115 D'Angelo G, Struman I, Martial J, Weiner RI: Activation of mitogen-activated protein kinases by vascular endothelial growth factor and basic fibroblast growth factor in capillary endothelial cells is inhibited by the antiangiogenic factor 16-kDa N-terminal fragment of prolactin. Proc Natl Acad Sci USA 1995;92:6374–6378.

116 Parenti A, Morbidelli L, Cui XL, Douglas JG, Hood JD, Granger HJ, Ledda F, Ziche M: Nitric oxide is an upstream signal of vascular endothelial growth factor-induced extracellular signal-regulated kinase1/2 activation in postcapillary endothelium. J Biol Chem 1998;273:4220–4226.

117 Shepard DR, Garcia JA: Toxicity associated with the long-term use of targeted therapies in patients with advanced renal cell carcinoma. Expert Rev Anticancer Ther 2009;9:795–805.

118 Hatzivassiliou G, Song K, Yen I, Brandhuber BJ, Anderson DJ, Alvarado R, Ludlam MJ, Stokoe D, Gloor SL, Vigers G, Morales T, Aliagas I, Liu B, Sideris S, Hoeflich KP, Jaiswal BS, Seshagiri S, Koeppen H, Belvin M, Friedman LS, Malek S: RAF inhibitors prime wild-type RAF to activate the MAPK pathway and enhance growth. Nature 2010;464:431–435.

119 Boehm S, Rothermundt C, Hess D, Joerger M: Antiangiogenic drugs in oncology: a focus on drug safety and the elderly – A mini-review. Gerontology 2010;56:303–309.

120 Ansari J, Glaholm J, McMenemin R, James ND, Hussain SA: Recent advances and future directions in the management of metastatic renal cell carcinoma. Anticancer Agents Med Chem 2010;10:225–235.

121 Lefevre G, Glotin AL, Calipel A, Mouriaux F, Tran T, Kherrouche Z, Maurage CA, Auclair C, Mascarelli F: Roles of stem cell factor/c-Kit and effects of Glivec/STI571 in human uveal melanoma cell tumorigenesis. J Biol Chem 2004;279:31769–31779.

122 Shah NP, Tran C, Lee FY, Chen P, Norris D, Sawyers CL: Overriding imatinib resistance with a novel ABL kinase inhibitor. Science 2004;305:399–401.

123 Lindauer M, Hochhaus A: Dasatinib. Recent Results Cancer Res 2010;184:83–102.

Martine J. Jager
Department of Ophthalmology
Leiden University Medical Centre (LUMC)
NL–2300 RC Leiden (The Netherlands)
E-Mail m.j.jager@lumc.nl

Jager MJ, Desjardins L, Kivelä T, Damato BE (eds): Current Concepts in Uveal Melanoma.
Dev Ophthalmol. Basel, Karger, 2012, vol 49, pp 137–149

Immunotherapy of Uveal Melanoma

Jacobus J. Bosch

Department of Internal Medicine 5, Hematology and Medical Oncology, University Hospital Erlangen, Erlangen, Germany

Abstract

Uveal melanoma is a malignancy with exceptional features for treatment with immunotherapy. Primary uveal melanoma can be treated with a variety of therapies that may limit the growth of the primary tumor in the eye and partially preserve vision. However, none of these treatment modalities prevents the development of metastases, which predominantly arise in the liver and universally remain fatal. Novel therapies are being explored for their effectiveness against uveal melanoma metastases and immunotherapy may be a potential option as an alternative or adjunctive treatment, even in the prophylactic setting. Uveal melanoma may be particularly responsive to T-cell-based immunotherapy because it originates in the immune-privileged eye. The localization of the primary tumor in the immune-privileged eye excuses the tumor cells from continuous immunological pressure. This may render primary uveal melanoma more immunogenic than tumor cells from non-privileged sites and allow expression of novel tumor antigens to which the patient's endogenous T cell repertoire is not tolerized. The clinical and genetic differences between cutaneous and uveal melanomas underscore the need for immunotherapy specifically designed for uveal melanoma patients. In this review, the current developments in the field of immunotherapy for uveal melanoma are discussed, with a special emphasis on T-cell-based strategies.

Uveal melanoma is the most common cancer of the eye and it metastasizes in up to 50% of patients with large tumors. The primary tumor in the eye can be treated successfully with a variety of therapies. None of these therapies, however, prevents the development of metastases. In general, patients diagnosed with primary uveal melanoma do not present with detectable metastatic disease, and have a long disease-free interval before metastases become clinically evident [1]. Metastatic disease occurs predominantly in the liver and is associated with poor survival (median 2–9 months) [1, 2]. Strategies have been developed to control metastatic disease; however, success is limited and metastatic uveal melanoma remains universally fatal. Novel therapies are being explored for their effectiveness against uveal melanoma metastases and immunotherapy may be a potential option, even in the prophylactic setting, to prevent metastatic disease. Since uveal melanomas develop in the immune-privileged environment of the eye, these tumors may express novel and immunogenic tumor antigens to which the patient's endogenous T cells are not tolerized. Uveal melanoma may therefore be particularly responsive to T-cell-based immunotherapy.

Protective Antitumor Immune Responses Capitalize on Adaptive Immunity by Tumor-Specific T Cells

Since the early stages of immunology, immune responses to cancer have intrigued immunologists such as Paul Ehrlich. Laboratory data and clinical observations have led to the development of a variety of cancer immunotherapies. Simultaneously, a growing body of evidence suggests that immunosurveillance continuously controls tumor development [3]. This immunological pressure may also induce mechanisms that enable tumor cells to suppress or evade immune recognition and lead to the development of tumor escape variants. Therefore, the challenge for immunotherapy is to induce long-lasting, protective antitumor immunity and overcome tumor escape from immune recognition.

The immune system interacts with tumor cells via the innate and adaptive arms of the immune response. Studies on immunotherapy for melanoma have mostly focused on adaptive immunity by antigen-specific T lymphocytes due to their antigen specificity and memory [4]. Specificity is required to prevent destruction of normal tissues and memory is required to prevent recurrences of primary and metastatic tumors. There are two major T lymphocyte populations, CD8+ and CD4+ T cells, which recognize distinct antigens and display distinct effector functions.

CD8+ Cytotoxic T Cells

CD8+ cytotoxic T cells (CTL) possess T cell receptors (TCR) that recognize small peptide antigens presented by major histocompatibility class I molecules, in humans designated human leukocyte antigen (HLA) class I. CTL activation proceeds through clonal expansion of precursor CD8+ T cells, differentiation into cells that acquire cytolytic capacity, ending with fully mature CTL that are highly specific. CTL are resistant to their own cytolytic factors and therefore able to survive and proceed rapidly to kill another tumor target cell. By this recycling mechanism, a small number of CTL can eliminate a much larger tumor burden. For these reasons, T-cell-based immunotherapies have predominantly focused on CD8+ T cells [4].

CD4+ Helper T Cells

In our efforts to develop an immunotherapy for uveal melanoma, we have focused on the activation of CD4+ T lymphocytes which have long been recognized as critical for optimal CD8+ T-cell-mediated immunity [5], either through their classical role as 'helper' T cells that provide cytokine support for CD8+ T cells [6], or by their induction of CD40 expression on dendritic cells (DC) ('licensing'), which in turn activate CD8+ T cells [7]. CD4+ T helper (Th) cells are also essential for generating CD8+ T memory cells and for preventing CD8+ T cells from being tolerized [8].

Historically, CD4+ Th cells were classified as type 1 CD4+ T cells (Th1) or type 2 CD4+ T cells (Th2) based on their cytokine profile [9]. Th1 cells produce interleukin-2 (IL-2) and Interferon-γ (IFN-γ), and facilitate tumor regression. Th2 cells produce IL-4 and IL-10, and favor tumor progression. In recent years, this Th1/Th2 paradigm has shifted with the identification of Th17 cells that produce IL-17 and play a role in the induction of autoimmunity [10]. Several novel subpopulations of CD4+ Th cells with distinct effector functions have subsequently been classified based on their cytokine profile and include Th3, Th9, Th22, T follicular helper cells, Tr cells, induced T regulatory cells and natural T regulatory cells [11].

CD4+ Th cells are restricted by HLA class II molecules. Studies on CD4+ Th cells have been facilitated by the identification of HLA class-II-restricted tumor antigen epitopes and novel cytokines. Conceptually, it may seem challenging to envision a direct effector function of CD4+ Th cells in recognition of tumor cells as most tumor cells do not express HLA class II. However, since CD4+ Th cells mediate their effects by interacting with CD8+ T cells and/or DC, once activated, they do not need to directly react with tumor cells [12].

Tumor-Infiltrating Lymphocytes

Similar to cutaneous melanomas, tumor-infiltrating lymphocytes (TIL) have been found in uveal melanomas [13–19]. CD8+ T cells from the peripheral blood of uveal melanoma patients, or TIL isolated from primary uveal melanomas are capable of lysing human uveal melanoma cells in vitro [16, 20, 21]. Of interest, Nitta et al. [13] documented in a small sample size that TIL in uveal melanomas express a restricted TCR repertoire predominantly consisting of Vα7. This restricted TCR gene usage in uveal melanoma TIL suggests that a specific antigen is targeted.

Uveal Melanoma-Induced Immune Suppression Is a Hurdle to Overcome

The loss or downregulation of HLA class I is an important immune escape mechanism that is exploited by tumor cells to avoid T cell recognition and promote tumor progression. In contrast to the majority of tumors, in uveal melanoma HLA class I expression is upregulated during progression to metastatic disease and correlates with a poor prognosis [22, 23]. This increased HLA class I expression implies that uveal melanoma metastases would be good targets for T cells; but are they really?

Direct Immunosuppression

A number of directly immunosuppressive properties of uveal melanomas have been identified and include local secretion of Tumor Growth Factor β (TGF-β) [24], and IFN-γ-mediated induction of the enzyme indoleamine 2,3 dioxygenase that depletes the local environment of tryptophan necessary for T cell clonal expansion, proliferation and survival [25]. In addition, constitutive or IFN-γ-induced expression of programmed death ligand-1 by primary uveal melanomas and its metastases inhibits T cell activation via binding to program death-1 on the T cell [26]. Uveal melanoma cells are resistant to Fas ligand-induced apoptosis by CTL, despite their expression of both Fas and Fas ligand [27]. Furthermore, IFN-γ-stimulated uveal melanoma cells become resistant to perforin-mediated cytolysis by MHC-class-I-restricted, cytolytic CD8+ T cells [28]. Of note, although IFN-γ is an important cytokine that supports T cell activation, it seems to have two faces in immunomodulation of uveal melanoma.

Indirect Immunosuppression

Uveal melanoma cells can indirectly inhibit antitumor immune responses via induction of immunosuppressive lymphoid and myeloid cell populations in the tumor microenvironment. Tumor-infiltrating T regulatory cells, which are predominantly CD4+ FOXP3+ T lymphocytes that produce TGF-β, have been observed in primary uveal melanoma tissue [29, 30]. Myeloid-derived suppressor cells are a heterogenous, immature population of myeloid cells that suppress immunity via a variety of mechanisms [31]. For example, myeloid-derived suppressor cells inhibit T cell activation via downregulation of the TCR-associated CD3 ζ-chain. In uveal melanoma patients, increased levels of myeloid-derived suppressor cells and downregulation of the CD3 ζ-chain on T cells in the peripheral blood [32] as well as tumor tissue [19] have been observed. The presence of macrophages in primary uveal melanoma stroma is correlated with a poor prognosis [33] and the majority of these tumor-infiltrating macrophages are of the M2 tumor-promoting, T-cell-suppressing subtype [34]. Collectively, these data indicate that immunosuppressive properties of uveal melanoma need to be targeted in order to successfully activate uveal melanoma-specific T cells.

Adapting Immunotherapy of Cutaneous Melanoma for Treatment of Uveal Melanoma

The shared origin of cutaneous and uveal melanoma suggests that immunotherapy developed for cutaneous melanoma would also be effective for uveal melanoma [35]. In general, uveal melanoma patients have been included in large cutaneous melanoma trials. However, anecdotal reports on small numbers of metastatic uveal melanoma patients included in these trials indicate that immunotherapy developed for cutaneous melanoma is unlikely to be effective. This could be due to the clinical and genetic differences between cutaneous and uveal melanomas.

Somatic mutations in the heterotrimeric G-protein α-subunit *GNAQ,* and its related gene *GNA11,* were recently reported to be frequently found in uveal and absent in cutaneous melanoma [36, 37]. The cytogenetic alterations monosomy 3, trisomy 8q and to some extent loss of 6q are characteristic for uveal melanoma [38]. These chromosomes may harbor somatically mutated genes important to disease progression, such as the recently identified somatic mutations in the gene encoding BRCA1-associated protein 1 *(BAP1)* on chromosome 3p21.1 [39]. Components that form the signal transduction pathway downstream of mutated genes, such as *GNAQ, GNA11* and *BAP1,* or other somatically mutated genes could reveal potential tumor antigen epitopes to which the patient's endogenous T cell repertoire may not be tolerized.

These observations emphasize the need for immunotherapy specifically designed for uveal melanoma patients. The existing immunotherapy regimens for cutaneous melanoma are not dismissed, but should strategically be reconsidered for a different mode of application, combination or timing. Two forms of immunotherapy for melanoma have been used: (1) approaches that boost the T cell response, including active immunization, and (2) adoptive cellular therapy.

Whole Cell-Based Vaccines

Whole tumor cell vaccines are based on the premise that the tumor cell can function as a nonprofessional antigen-presenting cell. Despite many regulatory challenges, the use of tumor cells in whole cell-based vaccine approaches is still considered attractive. It is favored since tumor cells generate unique antigens that are processed and presented differently in professional antigen-presenting cells. It is challenged for the predominantly immunosuppressive phenotype of tumor cells. Mitchell et al. [40] reported regression of a large primary choroidal melanoma after treatment with a whole cell-based vaccine. The vaccine, called Melacine, consisted of lysates of two human cutaneous melanoma cell lines administered with immunologic adjuvant Detox.

In vitro studies using unmodified uveal melanoma cells as priming agents for CD8+ T cells have given equivocal results. Verbik et al. [41] reported that uveal melanoma cells inhibited the activation of CD8+ T cells, and attributed the inhibition to the lack of costimulation and HLA class II expression. Two other studies reported the activation of CD8+ T cells; however, most of these responses were not restricted to HLA class I, and it was unclear if the activated CD8+ T cells were responding to uveal melanoma tumor antigens or to alloantigens [20, 42]. Genetic modification of uveal melanoma cells to express the costimulatory molecule CD80 resulted in slightly better T cell stimulation and enhanced surrogate antigen-presenting cell function [43, 44], but lacked HLA class II expression required for direct CD4+ Th cell activation.

Cell-based vaccines were created that consist of uveal melanoma cells that constitutively express HLA class I molecules, do not constitutively express HLA class II molecules, and are genetically modified to express CD80 costimulatory molecules and HLA class II alleles that are syngeneic to the recipient [45]. For uveal melanoma, these vaccines are called MHC II uveal melanoma vaccines (Oculovax™) and were specifically designed to activate uveal melanoma-specific, HLA-DR-restricted CD4+ T cells and thereby generate protective antitumor immunity and immune memory in uveal melanoma patients who are at high risk of developing metastatic disease [46]. Our studies demonstrated that MHC II uveal melanoma vaccines efficiently activate tumor-reactive, IFN-γ-secreting, HLA-DR-restricted CD4+ T cells as well as tumor-specific, cytotoxic CD8+ T cells from healthy donors and uveal melanoma patients [46, 47]. Vaccines prepared from individual patients' primary uveal melanoma cells activated CD4+ and CD8+ T cells that cross-reacted with aggressive primary and metastatic tumor cells derived from other uveal melanoma patients. Furthermore, expression of CD80 costimulatory molecules prevents constitutive and IFN-γ-induced uveal melanoma expression of programmed death ligand-1, and thereby overcomes T cell suppression [48].

Due to HLA polymorphism, an MHC class II uveal melanoma vaccine is likely to be partially HLA class I allogeneic to the recipient. However, HLA class I matching will be feasible, since the HLA-A2 allele is expressed by approximately 50% of uveal melanoma patients [49]. Interestingly, CD8+ T cells activated by the vaccines are cytolytic for HLA class I mismatched, as well as syngeneic, uveal melanoma targets [47]. Kan-Mitchell and colleagues [20, 42] also found that CD8+ T cells from patients with uveal melanoma recognized targets independent of HLA class I genotype. Whether uveal melanomas are unusual in inducing this type of allogeneic cytotoxicity and how the activated CD8+ T cells recognize HLA class I mismatched targets remains unclear.

MHC class II uveal melanoma vaccines must be matched to the HLA class II haplotype of the uveal melanoma patient for only one HLA class II allele. Since HLA class II alleles are not as polymorphic as HLA class I alleles and matching for MHC class I alleles is not necessary, it will not be necessary to generate a specific vaccine for each patient. Instead, a 'cocktail vaccine' consisting of a pool of 4–6 individual primary uveal melanoma cell lines each transduced with CD80 and an HLA-DR allele shared with the patient may be the most effective and feasible approach for adapting the MHC II uveal melanoma vaccines for clinical use [46].

A patient with advanced metastatic uveal melanoma has been treated on a compassionate basis with an MHC II uveal melanoma vaccine at the Hadassah University Hospital, Jerusalem, Israel. This patient survived well beyond the expected survival time and had shown no signs of adverse effects or autoimmunity [Prof. Dr. S. Slavin, pers. commun.]. Although statistically significant conclusions cannot be made from a single patient, immunization of the patient with the MHC II uveal melanoma vaccine may have elicited a detectable T cell response.

Dendritic Cell- and Peptide-Based Vaccines

Therapeutic vaccination with tumor antigen-derived peptides is aimed at the activation of the patient's endogenous tumor-specific T cells. These peptides are typically loaded and presented by DC that are critical for the modulation of effector CD4+ and CD8+ T cell responses. Provided they are properly activated, DC superiorly facilitate T cell priming and boosting, and are therefore frequently included in peptide-based and other immunotherapy approaches [50].

Uveal melanomas may be good targets for peptide-based vaccine therapy, because tumor cells express tumor antigens, including the melanoma-specific MAGE antigens MAGE-1, -2, and -3, the melanoma-associated antigens gp100 and tyrosinase [51–54], which are known T cell targets in cutaneous melanoma. Since normal choroidal melanocytes and retinal pigment epithelial cells do not express MAGE antigens [52], and immunization of cutaneous melanoma patients with MAGE peptides has not produced significant autoimmune problems, targeting these antigens is therapeutically feasible.

Several clinical trials using peptide vaccination in patients with metastatic uveal melanoma are ongoing or have been completed (ClinicalTrails.gov). Valmori et al. [55] observed a successful CD8+ T cell response of a uveal melanoma patient to a vaccine composed of 4 different tumor antigen-derived peptides administered simultaneously in incomplete Freund's adjuvant. Two of the patient's uveal melanoma lymph node metastases expressed Melan-A, tyrosinase and gp100, but not NY-ESO-1. After vaccination, the CD8+ T cell response was readily detected ex vivo against the peptide NY-ESO-$1_{157–165}$, and at a later time point against the peptide analog Melan-$A_{26–35\ A27L}$, and the peptide $tyrosinase_{368–376}$. No detectable CD8+ T cell response was observed against the fourth peptide, $gp100_{457–466}$. A rapid decline in the vaccine-induced CD8+ T cell responses was observed after the initial response; however, subsequent boosting with further peptide injections restimulated the vaccine response.

In general, despite robust vaccine-specific immune responses as measured by ex vivo laboratory techniques, whole cell-based, peptide and/or DC vaccination have shown infrequent objective tumor regression [56]. Therefore, boosting of the in vivo T cell response by cancer vaccines need to be further improved. For example, tumor antigen presentation is optimized by transfection of DC with tumor antigen-encoding mRNA or total mRNA isolated from melanoma cells [57]. In addition, the use of longer peptides comprising more CD8+, or both CD4+ and CD8+ T cell epitopes has shown a more efficient activation of tumor-specific T cells [58]. Alternatively, approaches beyond active immunization are being developed, such as blockade of the T cell immunomodulatory molecule CTLA-4 by fully humanized antibodies (MDX-010; ipililumab) in conjunction with peptide vaccination [59].

Adoptive Cellular Therapy

Adoptive cellular therapy (ACT) for melanoma is a form of immunotherapy that is based on the ex vivo activation of tumor-specific T cells and reinfusion of effector T cells into melanoma patients [60]. Incidentally, uveal melanoma patients have been treated in ACT trials for cutaneous melanoma. One of the first ACT for metastatic melanoma consisted of the systemic infusion of autologous lymphocytes activated in vitro combined with IL-2 [61]. This approach was extended to the use of TIL isolated from tumor biopsies, expanded in vitro, and reinfused systemically into the patient. Subsequent modifications of these adoptive cell transfer approaches included preconditioning of the patient with nonmyeloablative and lymphodepleting chemotherapy regimens [62]. The primary objective of preconditioning is considered to be the removal of potential immunosuppressive regulatory T cells and to increase the levels of T cell growth-promoting factors [63]. To gain specificity, autologous tumor-derived lymphocytes were further enriched for CD8+ T cells or tumor-specific CD8+ T cell clones before reinfusion [64, 65]. More recently, ACT has been expanded to the use of CD4+ Th cells and T cells that are genetically modified to express TCR of known high avidity [66], or chimeric tumor antigen-specific receptors (T bodies) that target surface antigens independently of HLA class I. Hunder et al. [67] reported on a patient with refractory metastatic melanoma who developed a clinical remission after infusions with in vitro isolated and expanded CD4+ Th cell clones, specific for the NY-ESO-1 antigen. This clinical observation in a single patient underscores the potential of CD4+ Th cells for ACT.

Although ACT is primarily aimed at inducing protective, systemic immune responses against metastases, it could have an effect on primary uveal melanomas as well. Some metastatic melanoma patients who received ACT and showed dramatic clinical responses at the same time developed autoimmunity in the form of vitiligo and uveitis [68, 69]. The observation of uveitis in these patients is particularly intriguing, since it suggests that ocular immune privilege is overcome and recognition of melanoma antigens occurs in the uveal tract. Whether similar responses could lead to intraocular tumor rejection remains to be seen.

Therapeutic and Prophylactic Immunotherapy for Uveal Melanoma

Immunotherapy for uveal melanoma could be applicable during two stages of the disease: (1) in the (neo)adjuvant setting, i.e. at the time of metastatic disease, and (2)

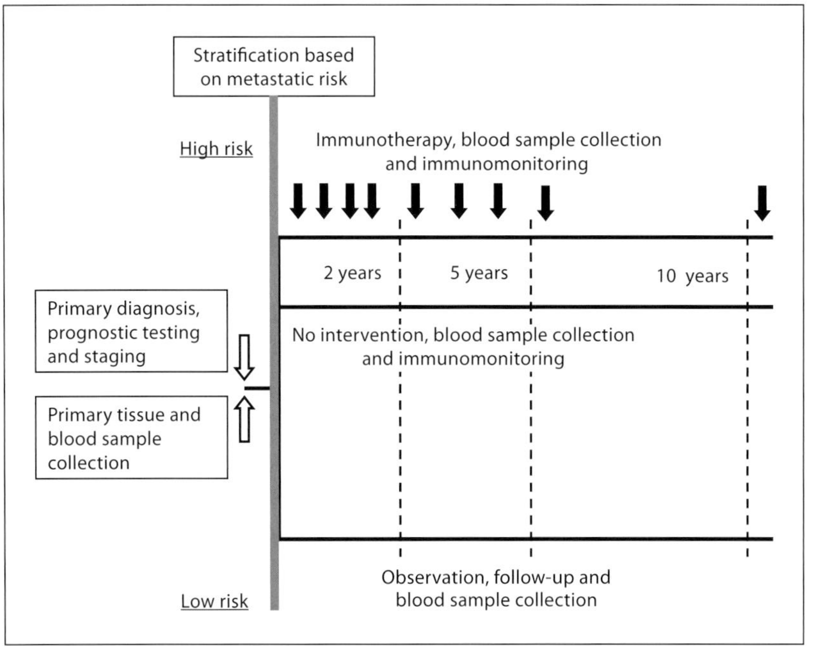

Fig. 1. Study design of immunotherapy for high-risk uveal melanoma patients. Immunotherapy for uveal melanoma could be tested in the prophylactic setting, i.e. at the time of primary diagnosis and during follow-up, when no metastatic disease is clinically evident. In an open-label, nonrandomized, intervention study, patients are stratified according to metastatic risk (gray bar; low risk and high risk) based on prognostic test results at primary diagnosis. High-risk uveal melanoma patients are further stratified into a treatment or observation arm. As depicted by black arrows, immunotherapeutic intervention is performed at regular intervals during a 2-, 5- and 10-year time frame, complemented at a minimum with semester and annual follow-up. At all time points, uveal melanoma tissue and/or blood sample collection is crucial for the development of immunotherapeutics and appropriate immunomonitoring. Interventions can include vaccination and/or ACT, among other forms of immunotherapy. Outcome measures include immunological and clinical response, overall and progression-free survival.

in the prophylactic setting, i.e. at the time of primary diagnosis and during follow-up, when no metastatic disease is clinically evident (fig. 1).

Tumors that contain monosomy 3 have an increased risk of developing metastases [70]. In addition, based on microarray studies, primary uveal melanomas can be divided into two molecular classes, class I and class II, of which primary tumors that express certain epithelial characteristics (class II tumors) metastasize at much higher frequencies [71]. If cytogenetic and microarray studies identify classifications of the primary tumor that are sufficiently prognostic of metastatic disease, then immunotherapy could also be used as a prophylactic treatment for patients who are at high risk of developing metastatic disease.

Novel clinical trials are currently active in which high-risk, monosomy 3, HLA-A2^{+} uveal melanoma patients are vaccinated with either melanoma antigen-derived peptides or melanoma antigen-encoding mRNA-transfected DC in the prophylactic setting (ClinicalTrails.gov). These immunotherapy trials are unique in that they vaccinate at early time points in the disease when there is no clinical evidence of metastatic disease. Furthermore, at this stage of the disease, bulky metastatic tumor load is absent and patients are probably not immunosuppressed, hence more responsive to any form of (active) immunotherapy. Long-term follow-up of these patients including appropriate immunomonitoring will be very interesting and may support the case for immunotherapy at early stages of malignant disease in general. Most importantly, it will hopefully provide survival benefit for this universally fatal disease.

Acknowledgements

The author thanks Dr. C.J. Voskens for critical reading of the manuscript and all the individuals who continue to support his work on immunotherapy of uveal melanoma, including Dr. M.J. Jager, Dr. P.W. Chen, Prof. Dr. B.R. Ksander, Prof. Dr. S. Ostrand-Rosenberg, Prof. Dr. C.J.M. Melief and Prof. Dr. A. Mackensen. Financial support was provided by a postdoctoral fellowship from Fight for Sight, Inc., New York, NY, USA, the following Dutch foundations: Rotterdamse Vereniging Blindenbelangen, Stichting Blindenhulp, Stichting Blinden-Penning, Stichting Dondersfonds, Stichting Nelly Reef Fund, Gratama Stichting, Stichting Admiraal van Kinsbergen Fonds, and Foundation 'De Drie Lichten', and the following German foundations: IZKF and ELAN Fonds of the University Hospital Erlangen.

Disclosure Statement

The author is co-holder of a patent concerning MHC II uveal melanoma vaccines (Oculovax).

References

1 Diener-West M, Reynolds SM, Agugliaro DJ, Caldwell R, Cumming K, Earle JD, Hawkins BS, Hayman JA, Jaiyesimi I, Jampol LM, Kirkwood JM, Koh WJ, Robertson DM, Shaw JM, Straatsma BR, Thoma J: Development of metastatic disease after enrollment in the COMS trials for treatment of choroidal melanoma: Collaborative Ocular Melanoma Study Group Report No 26. Arch Ophthalmol 2005;123:1639–1643.

2 Augsburger JJ, Correa ZM, Shaikh AH: Effectiveness of treatments for metastatic uveal melanoma. Am J Ophthalmol 2009;148:119–127.

3 Dunn GP, Old LJ, Schreiber RD: The immunobiology of cancer immunosurveillance and immunoediting. Immunity 2004;21:137–148.

4 Mocellin S, Mandruzzato S, Bronte V, Lise M, Nitti D: Vaccines for solid tumours. Part 1. Lancet Oncol 2004;5:681–689.

5 Keene JA, Forman J: Helper activity is required for the in vivo generation of cytotoxic T lymphocytes. J Exp Med 1982;155:768–782.

6 Bennett SR, Carbone FR, Karamalis F, Miller JF, Heath WR: Induction of a CD8+ cytotoxic T lymphocyte response by cross-priming requires cognate CD4+ T cell help. J Exp Med 1997;186:65–70.

7 Schoenberger SP, Toes RE, van der Voort EI, Offringa R, Melief CJ: T-cell help for cytotoxic T lymphocytes is mediated by CD40-CD40L interactions. Nature 1998;393:480–483.

8 Shedlock DJ, Shen H: Requirement for CD4 T cell help in generating functional CD8 T cell memory. Science 2003;300:337–339.
9 Mosmann TR, Cherwinski H, Bond MW, Giedlin MA, Coffman RL: Two types of murine helper T cell clone. 1. Definition according to profiles of lymphokine activities and secreted proteins. J Immunol 1986;136:2348–2357.
10 Bettelli E, Korn T, Oukka M, Kuchroo VK: Induction and effector functions of T(H)17 cells. Nature 2008; 453:1051–1057.
11 Kennedy R, Celis E: Multiple roles for CD4+ T cells in anti-tumor immune responses. Immunol Rev 2008;222:129–144.
12 Ossendorp F, Mengede E, Camps M, Filius R, Melief CJ: Specific T helper cell requirement for optimal induction of cytotoxic T lymphocytes against major histocompatibility complex class II negative tumors. J Exp Med 1998;187:693–702.
13 Nitta T, Oksenberg JR, Rao NA, Steinman L: Predominant expression of T cell receptor V alpha 7 in tumor-infiltrating lymphocytes of uveal melanoma. Science 1990;249:672–674.
14 de la Cruz PO Jr, Specht CS, McLean IW: Lymphocytic infiltration in uveal malignant melanoma. Cancer 1990;65:112–115.
15 Durie FH, Campbell AM, Lee WR, Damato BE: Analysis of lymphocytic infiltration in uveal melanoma. Invest Ophthalmol Vis Sci 1990;31:2106–2110.
16 Ksander BR, Rubsamen PE, Olsen KR, Cousins SW, Streilein JW: Studies of tumor-infiltrating lymphocytes from a human choroidal melanoma. Invest Ophthalmol Vis Sci 1991;32:3198–3208.
17 Meecham WJ, Char DH, Kaleta-Michaels S: Infiltrating lymphocytes and antigen expression in uveal melanoma. Ophthalmic Res 1992;24:20–26.
18 de Waard-Siebinga I, Hilders CG, Hansen BE, van Delft JL, Jager MJ: HLA expression and tumor-infiltrating immune cells in uveal melanoma. Graefes Arch Clin Exp Ophthalmol 1996;234:34–42.
19 Staibano S, Mascolo M, Tranfa F, Salvatore G, Mignogna C, Bufo P, Nugnes L, Bonavolonta G, De Rosa G: Tumor infiltrating lymphocytes in uveal melanoma: a link with clinical behavior? Int J Immunopathol Pharmacol 2006;19:171–179.
20 Kan-Mitchell J, Liggett PE, Harel W, Steinman L, Nitta T, Oksenberg JR, Posner MR, Mitchell MS: Lymphocytes cytotoxic to uveal and skin melanoma cells from peripheral blood of ocular melanoma patients. Cancer Immunol Immunother 1991;33: 333–340.
21 Ksander BR, Geer DC, Chen PW, Salgaller ML, Rubsamen P, Murray TG: Uveal melanomas contain antigenically specific and non-specific infiltrating lymphocytes. Curr Eye Res 1998;17:165–173.
22 Ericsson C, Seregard S, Bartolazzi A, Levitskaya E, Ferrone S, Kiessling R, Larsson O: Association of HLA class I and class II antigen expression and mortality in uveal melanoma. Invest Ophthalmol Vis Sci 2001;42:2153–2156.
23 Jager MJ, Hurks HM, Levitskaya J, Kiessling R: HLA expression in uveal melanoma: there is no rule without some exception. Hum Immunol 2002;63: 444–451.
24 Esser P, Grisanti S, Bartz-Schmidt K: TGF-beta in uveal melanoma. Microsc Res Tech 2001;52:396–400.
25 Chen PW, Mellon JK, Mayhew E, Wang S, He YG, Hogan N, Niederkorn JY: Uveal melanoma expression of indoleamine 2,3-deoxygenase: establishment of an immune privileged environment by tryptophan depletion. Exp Eye Res 2007;85:617–625.
26 Yang W, Chen PW, Li H, Alizadeh H, Niederkorn JY: PD-L1:PD-1 interaction contributes to the functional suppression of T-cell responses to human uveal melanoma cells in vitro. Invest Ophthalmol Vis Sci 2008;49:2518–2525.
27 Hallermalm K, De Geer A, Kiessling R, Levitsky V, Levitskaya J: Autocrine secretion of Fas ligand shields tumor cells from Fas-mediated killing by cytotoxic lymphocytes. Cancer Res 2004;64:6775–6782.
28 Hallermalm K, Seki K, De Geer A, Motyka B, Bleackley RC, Jager MJ, Froelich CJ, Kiessling R, Levitsky V, Levitskaya J: Modulation of the tumor cell phenotype by IFN-gamma results in resistance of uveal melanoma cells to granule-mediated lysis by cytotoxic lymphocytes. J Immunol 2008;180: 3766–3774.
29 Lagouros E, Salomao D, Thorland E, Hodge DO, Vile R, Pulido JS: Infiltrative T regulatory cells in enucleated uveal melanomas. Trans Am Ophthalmol Soc 2009;107:223–228.
30 Mougiakakos D, Johansson CC, Trocme E, All-Ericsson C, Economou MA, Larsson O, Seregard S, Kiessling R: Intratumoral forkhead box P3-positive regulatory T cells predict poor survival in cyclooxygenase-2-positive uveal melanoma. Cancer 2010;116:2224–2233.
31 Ostrand-Rosenberg S, Sinha P: Myeloid-derived suppressor cells: linking inflammation and cancer. J Immunol 2009;182:4499–4506.
32 McKenna KC, Beatty KM, Bilonick RA, Schoenfield L, Lathrop KL, Singh AD: Activated CD11b+ CD15+ granulocytes increase in the blood of patients with uveal melanoma. Invest Ophthalmol Vis Sci 2009;50:4295–4303.
33 Makitie T, Summanen P, Tarkkanen A, Kivela T: Tumor-infiltrating macrophages (CD68(+) cells) and prognosis in malignant uveal melanoma. Invest Ophthalmol Vis Sci 2001;42:1414–1421.

34 Bronkhorst IH, Ly LV, Jordanova ES, Vrolijk H, Versluis M, Luyten GP, Jager MJ: Detection of M2 macrophages in uveal melanoma and relation with survival. Invest Ophthalmol Vis Sci 2011;52:643–650.

35 van Dinten LC, Pul N, van Nieuwpoort AF, Out CJ, Jager MJ, van den Elsen PJ: Uveal and cutaneous melanoma: shared expression characteristics of melanoma-associated antigens. Invest Ophthalmol Vis Sci 2005;46:24–30.

36 Van Raamsdonk CD, Bezrookove V, Green G, Bauer J, Gaugler L, O'Brien JM, Simpson EM, Barsh GS, Bastian BC: Frequent somatic mutations of GNAQ in uveal melanoma and blue naevi. Nature 2009;457: 599–602.

37 Van Raamsdonk CD, Griewank KG, Crosby MB, Garrido MC, Vemula S, Wiesner T, Obenauf AC, Wackernagel W, Green G, Bouvier N, Sozen MM, Baimukanova G, Roy R, Heguy A, Dolgalev I, Khanin R, Busam K, Speicher MR, O'Brien J, Bastian BC: Mutations in GNA11 in uveal melanoma. N Engl J Med 2010;363:2191–2199.

38 Horsman DE, White VA: Cytogenetic analysis of uveal melanoma. Consistent occurrence of monosomy 3 and trisomy 8q. Cancer 1993;71:811–819.

39 Harbour JW, Onken MD, Roberson ED, Duan S, Cao L, Worley LA, Council ML, Matatall KA, Helms C, Bowcock AM: Frequent mutation of BAP1 in metastasizing uveal melanomas. Science 2010;330: 1410–1413.

40 Mitchell MS, Liggett PE, Green RL, Kan-Mitchell J, Murphree AL, Dean G, Spears L, Walonker F: Sustained regression of a primary choroidal melanoma under the influence of a therapeutic melanoma vaccine. J Clin Oncol 1994;12:396–401.

41 Verbik DJ, Murray TG, Tran JM, Ksander BR: Melanomas that develop within the eye inhibit lymphocyte proliferation. Int J Cancer 1997;73:470–478.

42 Huang XQ, Mitchell MS, Liggett PE, Murphree AL, Kan-Mitchell J: Non-fastidious, melanoma-specific CD8+ cytotoxic T lymphocytes from choroidal melanoma patients. Cancer Immunol Immunother 1994;38:399–405.

43 Uno T, Chen PW, Murray TG, Podack ER, Ksander BR: Gene transfer of the CD80 costimulatory molecule into ocular melanoma cells using a novel episomal vector. Invest Ophthalmol Vis Sci 1997;38: 2531–2539.

44 Carlring J, Shaif-Muthana M, Sisley K, Rennie IG, Murray AK: Apoptotic cell death in conjunction with CD80 costimulation confers uveal melanoma cells with the ability to induce immune responses. Immunology 2003;109:41–48.

45 Dissanayake SK, Thompson JA, Bosch JJ, Clements VK, Chen PW, Ksander BR, Ostrand-Rosenberg S: Activation of tumor-specific CD4(+) T lymphocytes by major histocompatibility complex class II tumor cell vaccines: a novel cell-based immunotherapy. Cancer Res 2004;64:1867–1874.

46 Bosch JJ, Thompson JA, Srivastava MK, Iheagwara UK, Murray TG, Lotem M, Ksander BR, Ostrand-Rosenberg S: MHC class II-transduced tumor cells originating in the immune-privileged eye prime and boost CD4(+) T lymphocytes that cross-react with primary and metastatic uveal melanoma cells. Cancer Res 2007;67:4499–4506.

47 Bosch JJ, Iheagwara UK, Reid S, Srivastava MK, Wolf J, Lotem M, Ksander BR, Ostrand-Rosenberg S: Uveal melanoma cell-based vaccines express MHC II molecules that traffic via the endocytic and secretory pathways and activate CD8(+) cytotoxic, tumor-specific T cells. Cancer Immunol Immunother 2010;59:103–112.

48 Haile ST, Bosch JJ, Agu N, Zeender A, Somasundaram P, Srivastava MK, Rödel S, Wolf J, Ksander BR, Ostrand-Rosenberg S: Tumor cell programmed death ligand-1-mediated T cell suppression is overcome by co-expression of CD80. Submitted.

49 Maat W, Haasnoot GW, Claas FH, Schalij-Delfos NE, Schreuder GM, Jager MJ: HLA Class I and II genotype in uveal melanoma: relation to occurrence and prognosis. Invest Ophthalmol Vis Sci 2006;47:3–6.

50 Melief CJ: Cancer immunotherapy by dendritic cells. Immunity 2008;29:372–383.

51 Mulcahy KA, Rimoldi D, Brasseur F, Rodgers S, Lienard D, Marchand M, Rennie IG, Murray AK, McIntyre CA, Platts KE, Leyvraz S, Boon T, Rees RC: Infrequent expression of the MAGE gene family in uveal melanomas. Int J Cancer 1996;66:738–742.

52 Chen PW, Murray TG, Salgaller ML, Ksander BR: Expression of MAGE genes in ocular melanoma cell lines. J Immunother 1997;20:265–275.

53 Chen PW, Murray TG, Uno T, Salgaller ML, Reddy R, Ksander BR: Expression of MAGE genes in ocular melanoma during progression from primary to metastatic disease. Clin Exp Metastasis 1997;15:509–518.

54 Luyten GP, van der Spek CW, Brand I, Sintnicolaas K, de Waard-Siebinga I, Jager MJ, de Jong PT, Schrier PI, Luider TM: Expression of MAGE, gp100 and tyrosinase genes in uveal melanoma cell lines. Melanoma Res 1998;8:11–16.

55 Valmori D, Dutoit V, Ayyoub M, Rimoldi D, Guillaume P, Lienard D, Lejeune F, Cerottini JC, Romero P, Speiser DE: Simultaneous CD8+ T cell responses to multiple tumor antigen epitopes in a multipeptide melanoma vaccine. Cancer Immun 2003;3:15.

56 Rosenberg SA, Yang JC, Restifo NP: Cancer immunotherapy: moving beyond current vaccines. Nat Med 2004;10:909–915.
57 Schaft N, Dorrie J, Thumann P, Beck VE, Muller I, Schultz ES, Kampgen E, Dieckmann D, Schuler G: Generation of an optimized polyvalent monocyte-derived dendritic cell vaccine by transfecting defined RNAs after rather than before maturation. J Immunol 2005;174:3087–3097.
58 Melief CJ, van der Burg SH: Immunotherapy of established (pre)malignant disease by synthetic long peptide vaccines. Nat Rev Cancer 2008;8:351–360.
59 Hodi FS, O'Day SJ, McDermott DF, Weber RW, Sosman JA, Haanen JB, Gonzalez R, Robert C, Schadendorf D, Hassel JC, Akerley W, van den Eertwegh AJ, Lutzky J, Lorigan P, Vaubel JM, Linette GP, Hogg D, Ottensmeier CH, Lebbe C, Peschel C, Quirt I, Clark JI, Wolchok JD, Weber JS, Tian J, Yellin MJ, Nichol GM, Hoos A, Urba WJ: Improved survival with ipilimumab in patients with metastatic melanoma. N Engl J Med 2010;363:711–723.
60 Rosenberg SA, Dudley ME: Adoptive cell therapy for the treatment of patients with metastatic melanoma. Curr Opin Immunol 2009;21:233–240.
61 Rosenberg SA, Lotze MT, Muul LM, Leitman S, Chang AE, Ettinghausen SE, Matory YL, Skibber JM, Shiloni E, Vetto JT, et al: Observations on the systemic administration of autologous lymphokine-activated killer cells and recombinant interleukin-2 to patients with metastatic cancer. N Engl J Med 1985;313:1485–1492.
62 Dudley ME, Wunderlich JR, Yang JC, Sherry RM, Topalian SL, Restifo NP, Royal RE, Kammula U, White DE, Mavroukakis SA, Rogers LJ, Gracia GJ, Jones SA, Mangiameli DP, Pelletier MM, Gea-Banacloche J, Robinson MR, Berman DM, Filie AC, Abati A, Rosenberg SA: Adoptive cell transfer therapy following non-myeloablative but lymphodepleting chemotherapy for the treatment of patients with refractory metastatic melanoma. J Clin Oncol 2005;23:2346–2357.
63 Gattinoni L, Finkelstein SE, Klebanoff CA, Antony PA, Palmer DC, Spiess PJ, Hwang LN, Yu Z, Wrzesinski C, Heimann DM, Surh CD, Rosenberg SA, Restifo NP: Removal of homeostatic cytokine sinks by lymphodepletion enhances the efficacy of adoptively transferred tumor-specific CD8+ T cells. J Exp Med 2005;202:907–912.
64 Yee C, Thompson JA, Byrd D, Riddell SR, Roche P, Celis E, Greenberg PD: Adoptive T cell therapy using antigen-specific CD8+ T cell clones for the treatment of patients with metastatic melanoma: in vivo persistence, migration, and antitumor effect of transferred T cells. Proc Natl Acad Sci USA 2002;99:16168–16173.
65 Mackensen A, Meidenbauer N, Vogl S, Laumer M, Berger J, Andreesen R: Phase I study of adoptive T-cell therapy using antigen-specific CD8+ T cells for the treatment of patients with metastatic melanoma. J Clin Oncol 2006;24:5060–5069.
66 Morgan RA, Dudley ME, Wunderlich JR, Hughes MS, Yang JC, Sherry RM, Royal RE, Topalian SL, Kammula US, Restifo NP, Zheng Z, Nahvi A, de Vries CR, Rogers-Freezer LJ, Mavroukakis SA, Rosenberg SA: Cancer regression in patients after transfer of genetically engineered lymphocytes. Science 2006;314:126–129.
67 Hunder NN, Wallen H, Cao J, Hendricks DW, Reilly JZ, Rodmyre R, Jungbluth A, Gnjatic S, Thompson JA, Yee C: Treatment of metastatic melanoma with autologous CD4+ T cells against NY-ESO-1. N Engl J Med 2008;358:2698–2703.
68 Yeh S, Karne NK, Kerkar SP, Heller CK, Palmer DC, Johnson LA, Li Z, Bishop RJ, Wong WT, Sherry RM, Yang JC, Dudley ME, Restifo NP, Rosenberg SA, Nussenblatt RB: Ocular and systemic autoimmunity after successful tumor-infiltrating lymphocyte immunotherapy for recurrent, metastatic melanoma. Ophthalmology 2009;116:981–989.e1.
69 Palmer DC, Chan CC, Gattinoni L, Wrzesinski C, Paulos CM, Hinrichs CS, Powell DJ Jr, Klebanoff CA, Finkelstein SE, Fariss RN, Yu Z, Nussenblatt RB, Rosenberg SA, Restifo NP: Effective tumor treatment targeting a melanoma/melanocyte-associated antigen triggers severe ocular autoimmunity. Proc Natl Acad Sci USA 2008;105:8061–8066.
70 Tschentscher F, Husing J, Holter T, Kruse E, Dresen IG, Jockel KH, Anastassiou G, Schilling H, Bornfeld N, Horsthemke B, Lohmann DR, Zeschnigk M: Tumor classification based on gene expression profiling shows that uveal melanomas with and without monosomy 3 represent two distinct entities. Cancer Res 2003;63:2578–2584.
71 Onken MD, Worley LA, Ehlers JP, Harbour JW: Gene expression profiling in uveal melanoma reveals two molecular classes and predicts metastatic death. Cancer Res 2004;64:7205–7209.

Jacobus J. Bosch, MD
Department of Internal Medicine 5, Hematology and Medical Oncology, University Hospital Erlangen
DE–91054 Erlangen (Germany)
Tel. +49 9131 8545122, E-Mail jacobus.bosch@uk-erlangen.de

Jager MJ, Desjardins L, Kivelä T, Damato BE (eds): Current Concepts in Uveal Melanoma.
Dev Ophthalmol. Basel, Karger, 2012, vol 49, pp 150–165

Genetic Determinants of Uveal Melanoma

Jérôme Couturier[a] · Simon Saule[b]

[a]Department of Genetics, Institut Curie – Hôpital, Paris, and [b]Research Center, Institut Curie, INSERM U1021, CNRS UMR3347, University Paris-Sud, Orsay, France

Abstract

Uveal melanoma (UM) arises from neural crest-derived melanocytes of the choroid and the ciliary body. About 50% of patients develop metastatic disease despite efficient control of the primary tumor. For about 15 years, cytogenetic and, recently, genome-wide analysis techniques have shown that UM can be classified into 2 genomic groups correlating with prognostic clinicopathologic features: class 1 tumors, with a low risk of metastases, typically characterized by a gain of the 6p chromosome arm, often associated with a gain of the distal part of the 8q chromosome arm, and class 2 tumors, with a high metastatic risk, presenting loss of the entire chromosome 3 and gain of the entire 8q, related to the formation of isochromosomes. Genome-wide expression profiling has proved to be a powerful tool for separating these 2 classes. However, despite advances in the genomic and prognostic characterization of UM, the knowledge of pathways deregulated in these tumors is just emerging and, in contrast to cutaneous melanoma, no major predisposing genes are known. Altered or deregulated genes are reviewed in this chapter. Inactivating mutations have recently been identified by exome sequencing in gene *BAP1*, mapping to 3p21.1, in class 2 tumors. Among other discriminant genes identified from genome-wide expression profiling, *PTP4A3*, mapping to 8q24.3, coding for a protein promoting cell migration, is highly overexpressed in class 2 tumors. The overall expression signature of class 2 tumors suggests they may originate from neuroectodermal stem cells.

Uveal melanoma (UM) is a rare neoplasm arising from melanocytes of the choroid or the ciliary body. It is markedly different to cutaneous melanoma in its genetics, genome profiles and tumorigenesis. One of its major clinical features is its high propensity to metastasize despite successful treatment of the primary tumor: about 50% of the patients develop metastases, in 90% of cases to the liver (for a review, see Damato [1]). It has been known for about 20 years that UM shows specific chromosome rearrangements, some of them, besides certain clinicopathological factors (e.g. age, tumor size, location, cell type, microcirculatory patterns, extrascleral invasion), being related to prognosis (for a review, see Kilic et al. [2]). More recently, with the advent of high-resolution genome and expression profiling, efforts have been made to identify candidate genes in order to better understand the natural history of the disease and to better assess the prognosis of patients [3, 4]. In this chapter, we review the present knowledge about genetic features of UM at the constitutional and somatic levels.

Predisposition to Uveal Melanoma

Rare families of UM and an excess of UM cases among relatives of affected patients have been reported. Familial forms, defined by at least 2 cases of this rare tumor in relatives, account for 0.6% of cases [5]. Genetic predisposition may also be suspected if an early onset occurs or in case of multifocal or bilateral tumors. In a study on 121 patients with UM, Abdel-Rahman et al. [6] showed that predisposition may also involve other cancers, thus the frequency of patients with cancer susceptibility may be higher than previously thought. Transmission could correspond to an autosomal dominant mode with incomplete penetrance [5]. Suspected by analogy with susceptibility to cutaneous melanoma, *CDKN2A,* with its variants *ARF* and *INK4,* and *CDK4* have been proposed as candidate genes, but a germ line mutation of *CDKN2A/INK4* was found in only 1 UM family, and no mutations of *CDKN2A/ARF* and *CDK4* are known [7]. A germ line mutation of *CDKN2A/INK4* has been identified in 1 patient from a series of 385 patients with a sporadic UM [8]. Another candidate gene is *BRCA2,* as 2 carriers of a mutation of this gene, from 2 independent families, were affected by UM [9]. In addition, in a study of 143 Ashkenazic Jewish patients with UM, 4 carried the 6174delT *BRCA2* mutation [10]. Recently, Buecher et al. [11] found no deleterious germ line mutations of *CDKN2A/INK4, CDKN2A/ARF, CDK4,* and *BRCA1/2* in 25 individuals with suspected predisposition to UM. In conclusion, these genes are not involved in the predisposition to UM in the majority of cases. This confirms that the genetic determinism of UM differs from that of cutaneous melanoma.

Genomic Abnormalities in Uveal Melanoma

Cytogenetics

It has been shown since the 1990s that UM usually presents nonrandom, relatively simple, cytogenetic abnormalities. These rearrangements, affecting mainly chromosomes 1, 3, 6, and 8, have first been identified by karyotype analyses, then by fluorescence in situ hybridization and chromosomal comparative genomic hybridization (CGH), performed on tumor material obtained from enucleations (for reviews, see Kilic et al. [2] and Harbour [4]). Among these abnormalities, monosomy 3 appears as an early event, present in 50–60% of tumors, which is often associated with isochromosome 8q. The isochromosome 8q is prone to malsegregate during mitoses and to multiply into 2, 3 or more copies, leading to a high level of 8q gain [2]. Monosomy 3 correlates with ciliary body involvement, epithelioid histology, and a poor outcome, whereas disomy 3 tumors are considered as rarely leading to metastatic disease (for a review, see Mudhar et al. [12]). The homogeneity of the presence of the abnormality in monosomy 3 tumors has been a matter of debate [13–15]. In our series, we never suspected a heterogeneous monosomy 3 in tumors with an abnormal CGH array profile. Tumors with a normal copy number of chromosome 3 usually present a gain of 6p, often related to isochromosome 6p. Additionally, other recurrent chromosome alterations, such as losses of 1p and 16q, have been described [2, 12]. Although cytogenetic analyses were very fruitful in the initial identification of characteristic chromosome abnormalities in UM, they have serious limitations. Karyotypic analyses are practically possible for large tumors obtained by enucleation and local resection only. Moreover, they are dependent on in vitro cell growth, so they provide results only for the most proliferative cases and may lead to biased conclusions regarding the real frequency of chromosome rearrangements. Another limitation is their low resolution, about 5 Mb. Interphase fluorescence in situ hybridization necessarily explores very limited regions of the genome, in a limited number of nuclei; it is costly and time consuming when analyzing several markers, and poorly reliable in this tumor type, as variations of 1 copy only for each marker should be detected. This has led to the increasing use of genome-wide molecular techniques.

Genome-Wide DNA Copy Number Profiling

Genomic Classification of Tumors

Genome and expression profiling on microarrays makes it now possible to analyze tumors that are characterized by combined imbalances with a much higher resolution and without the biases and limitations of cytogenetic analyses. CGH or single nucleotide polymorphism (SNP) arrays produce high-density genome-wide DNA copy number profiles. SNP arrays have the advantage of providing the allelic status of chromosome segments, which is especially useful for chromosome 3 in UM, for which isodisomy, in which 1 copy of chromosome 3 is lost and the other is duplicated, may occur [16]. The multiplex ligation-dependent probe amplification (MLPA) technique is also used for evaluating the status of the chromosome regions that have been found to be relevant in UM [17]. Analyses have mainly been carried out on samples of large tumors, obtained from enucleated eyes, but, more recently, fine-needle aspirates obtained from smaller tumors in vivo have also been analyzed [18, 19]. These genome-wide techniques have now provided a precise overview of chromosome imbalances in UM [18–22]. No true amplifications (high-level multiplication of a limited segment of the genome) have been described in these tumors. Hierarchical unsupervised clustering performed on the basis of chromosome imbalances shows that UM can be classified into 2 groups separated by the status of chromosome 3: class 1 tumors,

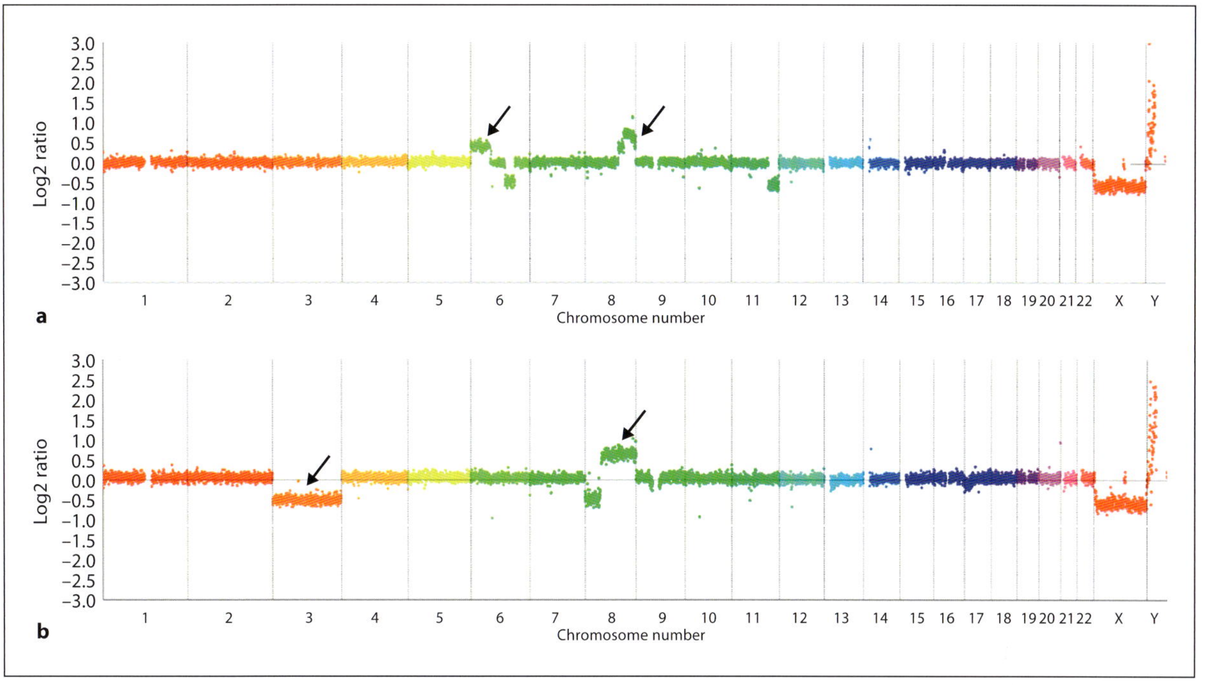

Fig. 1. Representative CGH array profiles of 2 UM; chip NimbleGen 72K. **a** Class 1 tumor; gains of 6p and distal 8q (arrows). **b** Class 2 tumor; loss of chromosome 3, gain of entire 8q (arrows).

with 2 copies, and class 2 tumors, with monosomy 3, representing about 40 and 60%, respectively, of tumors analyzed after enucleation. The class 1/class 2 nomenclature was defined first by Onken et al. [23] on the basis of expression profiling analysis. The proportion of class 2 tumors is only a bit lower in samples obtained by fine-needle aspiration, corresponding to smaller tumors: a mean of 46% in the studies of Young et al. [18] and McCannel et al. [19], and 50% in our series of 30 samples [unpubl. data]. In our CGH array analysis of 86 primary tumors obtained from enucleations [22], a complete monosomy 3 was present in 62% of the tumors, a partial or a total gain of the long arm of chromosome 8 in 89%, a gain of 6p in 57%, and a loss of 1p and 16q in 45 and 31%, respectively. The profiles of these tumors showed that class 1 tumors were mainly characterized by a gain of 6p and 8q, distal from band q21 (fig. 1a), and class 2 tumors by monosomy 3 and a gain of the entire 8q [22] (fig. 1b). Monosomy 3 or gain of 6p can be observed as sole imbalances, thus they are considered as a very early event. This suggests a bifurcated progression pathway in which monosomy 3 and gain of 6p are almost mutually exclusive alterations [3, 21, 24, 25]. In addition, Ehlers et al. [21] have shown that class 1 tumors exhibit low aneuploidy, whereas class 2 tumors have higher levels of aneuploidy.

Being present in both classes of UM, gain of 8q is the most frequent imbalance, but it occurs as either distal 8q gain, resulting from unbalanced translocations in class 1 tumors, or as whole 8q gain, related to isochromosomes in class 2 tumors, as shown by karyotype analyses. The gain of a whole 8q is often associated with loss of 8p; however, as seen when comparing the genomic profile and karyotype of a given tumor, 8p loss may be masked by the presence of 2 normal copies of chromosome 8, in addition to isochromosomes. These isochromosomes may accumulate in multiple copies leading to ratios considered as corresponding to amplifications of the whole 8q in genome profiles, but they are in fact only high-level overrepresentations of this chromosome arm. The gain of either a partial or a complete 8q appears to be a late event in the progression of the tumors.

Few data are available about genome-wide profiling of UM metastases. Our CGH array analysis of 63 liver metastases [22] showed that they clustered according to the same groups as primary tumors, but in different proportions: class 1 and class 2 represented 17 and 82% of metastases, respectively.

Chromosome 3

As the status of chromosome 3 is such a crucial marker in UM, several studies have been devoted to the precise detection of abnormalities of this chromosome. First, SNP array analyses have shown, by their ability to determine the allelic status of chromosome 3, that 3–5% of all UM exhibit isodisomy 3 [16, 22], having the same prognostic value as monosomy 3. This abnormality is copy-neutral, and not detectable by CGH array or MLPA. Secondly, microsatellite analysis, MLPA, or high-resolution genome-wide techniques have shown the existence of partial deletions of chromosome 3, either in the short or the long arm of the chromosome, or both [18, 19, 22, 26–28]. A recurrent minimal region of distal deletion has been found in 3p25.3 [22, 26–28] in primary UM, but in 2 liver metastases a minimal region of deletion has been found in 3p26.3 [22]. However, published cases with partial deletions of chromosome 3 are still scarce and analyzed by various techniques, and collaborative studies are warranted to collect data with homogeneous techniques, and to look for prognostic correlations.

Prognostic Correlations

From earlier cytogenetic studies we know that tumors with monosomy 3 and gain of 8q are associated with a high metastatic risk, whereas those with gain of 6p correlate with a good prognosis [2, 4, 12]. This has recently been confirmed and clarified by the use of high-resolution genome-wide CGH or SNP array techniques [15, 20–22] and MLPA [17], which have shown that class 1 and class 2 profiles were associated with low and high metastasis risk, respectively. SNP array appears as the most sensitive technique for these prognostic correlations thanks to its ability to detect chromosome 3 isodisomy [15]. It is worth noticing that a minority of class 1 tumors, probably about 15%, are able to metastasize, as confirmed by class 1 profiles observed in some liver metastases [22].

Candidate Genes

Despite the continuously increasing resolution of methods available for genome-wide DNA copy number analysis, minimal regions of imbalance remain relatively large in UM, and no candidate genes could be identified with certainty on this basis until now. *ASAP1 (DDEF1),* mapping to 8q24.21, has been proposed as a potential oncogene relevant to UM [29]. The same team identified in 4 rapidly metastasizing tumors a minimal region of deletion in 8p22-p12 containing a tumor suppressor gene, *LZTS1,* located in 8p21.3, which is downregulated. However, no mutations were found [30].

Molecular genetics analyses have identified a number of alterations reported as contributing to melanoma [3, 31]. These include constitutive activation of the MAP kinase pathway, for example via activating mutations in *BRAF* that are found in around 60% of cutaneous melanomas [32] and in 13% of UM [33], although only in subclones in this tumor. Mutations of the *GNAQ* gene occur in about 50% of UM [34], representing the most common known oncogenic mutation in this cancer. GNAQ is a heterotrimeric GTP-binding protein α subunit that couples G-protein-coupled receptor signaling to the RAF/MEK/ERK cascade and contributes to MAP kinase pathway activation. The presence of this mutation in tumors at all stages of malignant progression suggests that it is an early event in UM, but these mutations are not correlated with disease-free survival [35]. Recently, mutations in the *BAP1* gene, encoding a nuclear ubiquitin carboxyl-terminal hydrolase that binds to BRCA1 and BARD1 to form a tumor suppressor heterodimeric complex, have been identified by exome sequencing [36]. No correlation between *GNAQ* and *BAP1* was found, and *BAP1* mutations, observed in 84% of class 2 tumors, coincided with the onset of metastatic behavior.

Genome-Wide Expression Profiling

Gene expression profiling or transcriptome analysis involves the simultaneous measurement of messenger ribonucleic acid (mRNA) expression of multiple genes. The use of this technique together with a statistical approach has identified 2 major subgroups of UM, referred to as class 1 and class 2, related to genomic classes [23, 37]. The class 2 gene expression profile is closely associated with classical factors of poor prognosis, such as larger tumor size, epithelioid cytology, extravascular looping matrix patterns, and monosomy 3 [38–41], but the prognostic accuracy of the class 2 expression profile is higher than many of these factors taken individually or in combination. Optimization of transcriptome analysis for clinical applications has released a signature of as few as 12 discriminant genes that is able to classify UM tumors [14]. To test the efficiency of this classifier in metastasis prediction, we compared, in our own series of UM, primary tumors from patients who developed liver metastases within 3 years after enucleation with tumors from patients that did not develop metastases.

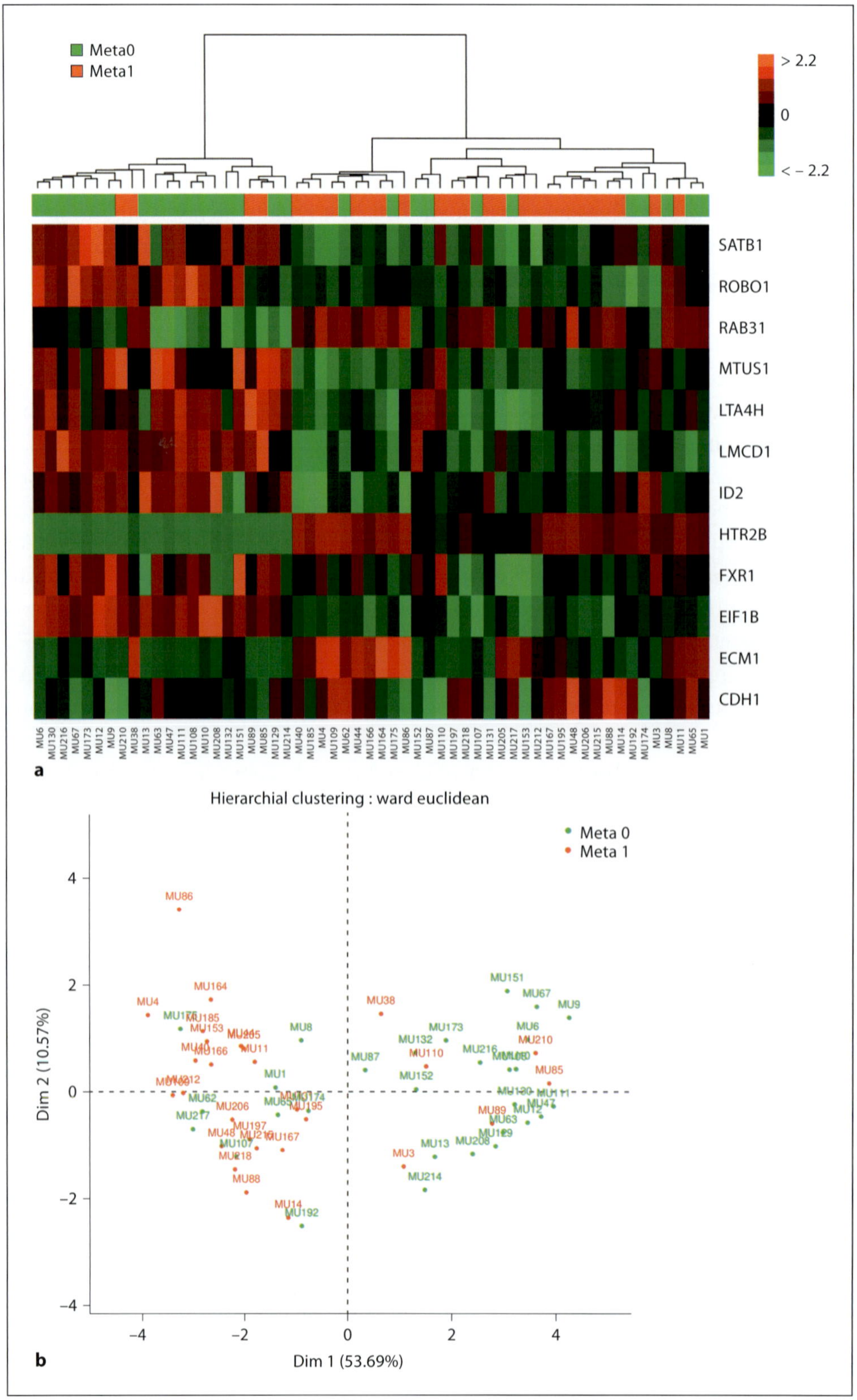

Meta0
Meta1
> 2.2
0
< – 2.2
SATB1
ROBO1
RAB31
MTUS1
LTA4H
LMCD1
ID2
HTR2B
FXR1
EIF1B
ECM1
CDH1
a
Hierarchial clustering : ward euclidean
Meta 0
Meta 1
Dim 2 (10.57%)
Dim 1 (53.69%)
b

The application of the 12-gene classifier [14] to the dataset of our study separates the tumors into 2 classes (fig. 2), but 21% of metastasizing tumors remain associated with the low-risk group (class 1), warranting further analyses of gene expression to identify genes more specifically associated with, and hopefully responsible for, metastasis [42]. Therefore, we compared the genes differentially expressed in patients with *meta1* tumors (patients who developed metastases within the 3-year follow-up) and *meta0* tumors (nonmetastasizing within the 3-year follow-up). Differentially expressed genes include the gene encoding protein tyrosine phosphatase type IV A member 3 *(PTP4A3),* which maps to 8q24.3 [43]. A high level of expression of this gene is highly predictive of metastases [44]. As mentioned before, DNA copy number analysis has shown that 8q is overrepresented in high-risk tumors; however, the expression profile of *PTP4A3* differed considerably from that of the neighboring genes on chromosome arm 8q, demonstrating that this gene was not simply a passenger gene. The immunodetection of PTP4A3 in tumor sections demonstrated a positive correlation between RNA and protein levels. Recently, PCBP1, an RNA-binding protein, has been demonstrated to be able to suppress the translation of PTP4A3 [45]. Since some *meta0* tumors expressed high levels of *PTP4A3* mRNA [42], a correlation study between PTP4A3 and PCBP1 immunostaining and metastasis occurrence remains to be performed.

Genetic Pathways in Uveal Melanoma

PTP4A3 Pathways

PTP4A3 is known to promote cell migration and invasion in cancer cells in vitro and in vivo [46–50]. UM cell lines stably producing wild-type, or mutant [47] PTP4A3 were established and time lapse video microscopy experiments revealed that cells producing the wild-type protein migrated faster, paused for shorter periods, and travelled further than did cells producing the mutant protein [42]. Cells producing the wild-type PTP4A3 had a significantly larger number of focal adhesions (labeled with antibodies against PTK2/FAK) covering a significantly smaller surface area than cells producing the mutant form of the phosphatase [42]. PTK2, located at 8q24.3, was also found to be differentially expressed between *meta0* and *meta1* tumors, and the RNA expression level of this gene is positively correlated with *PTP4A3* (Pearson coefficient: $R = 0.71$, $p < 0.001$). The formation and remodeling of focal contacts is a dynamic process regulated by protein tyrosine kinases and small GTPases of the Rho family [51], and modulated by PTP4A3.

Fig. 2. Hierarchical clustering (**a**) and principal component analysis (**b**) on 12 discriminating genes previously described [14]. Tumors are labeled according to their metastatic status (green: *meta0;* red: *meta1*).

Downregulation of PTP4A3 reduced metastasis spreading [52], and we confirmed with a naturally immunodeficient host chick embryo model that, 1 week after inoculation of PTP4A3, overexpressing UM cell lines were able to migrate and invade the avian blood system [42]. Thus, in addition to increasing cell migration, PTP4A3 may also be involved in the regulation of protease-encoding genes. The downstream targets of PTP4A3 are not yet well defined, and RNA studies correlated with PTP4A3 expression may indicate either transcriptional targets of PTP4A3, or factors that may act synergistically with PTP4A3 in the induction of metastases. Reciprocally, inversely correlated genes may antagonize PTP4A3 function. In primary tumors, *PTP4A3* expression correlated with the expression of several proteases (including ADAM10 and ADAM12, none of the gene coding for them being present in critically abnormal chromosomes) and inversely correlated with *TIMP3* protease inhibitor levels ($R = -0.57$, $p < 0.001$). *ADAM10* (a disintegrin and metalloproteinase 10) is involved in the ectodomain shedding of various adhesion molecules known to have important roles in the development of malignant melanoma and is upregulated in melanoma metastases [53]. *ADAM12* has been shown to digest extracellular matrix compounds, such as gelatin type IV collagen and fibronectin [54].

TCEB1 also positively correlated with *PTP4A3* ($R = 0.79$, $p < 0.001$) and is, as *PTP4A3*, located on chromosome 8, in 8q21.11. It is one of the 3 subunits of the elongin complex (SIII) that activates transcription, and promotes invasion of prostate cancer cells [55]. The invasion-promoting capability of TCEB1 could be due to the enhancing expression of invasion-related genes. TCEB1 is also part of the Von Hippel-Lindau complex, selectively targeting the HIF1a factor for degradation in the presence of oxygen [56], and playing a well-described role in melanoma [57]. The Von Hippel-Lindau complex is not the only protein TCEB1 binds to, and it remains to be tested whether PTP4A3 is able to interact or control the function of this transcription activator.

PTP4A3 has been reported to be able to regulate the C-terminal SRC kinase protein synthesis through the phosphorylation modulation of the elongation factor EIF2α [58]. In response to various environmental stresses, eukaryotic cells downregulate protein synthesis by phosphorylation of the α subunit of EIF2. EIF2 is composed of 3 nonidentical subunits, i.e. EIF2α, EIF2β, and EIF2γ [59]. Phosphorylation of EIF2α results in the inhibition of the guanine nucleotide exchange factor EIF2β, thereby reducing the rate of the GDP-to-GTP exchange that is required for EIF2 to carry out additional rounds of translation initiation. The eukaryotic EIF2α kinases share extensive homology within their catalytic domains, but have distinct regulatory domains allowing for different physiologic signals to regulate phosphorylation of EIF2α. EIF2AK1 mediates downregulation of protein synthesis by the phosphorylation of EIF2α. We observed that *EIF2AK1* is differentially expressed in *meta0* and *meta1* tumors and may therefore participate in the metastatic phenotype through protein synthesis regulation. C-terminal SRC kinase is a negative regulator of SRC. SRC activity is controlled by autophosphorylation of Tyr-416 in the kinase domain and through phosphorylation of Tyr-527 in the C-terminal tail. The latter phosphorylation event, catalyzed by C-terminal SRC kinase [60], represses SRC activity through an inhibitory intramolecular pTyr-527-SH2 interaction [61]. SRC kinase initiates a number of signal pathways responsible for increased cell growth and motility [62], including PTK2 from focal adhesions [63]. This is consistent with SRC being activated in a large number of malignancies and cancer metastases [64, 65], and potentially being an indirect target of PTP4A3.

ASAP1/DDEF1

Development and differentiation enhancing factor 1 *(DDEF1),* mapping to 8q24.21, also known as Arf-GAP containing SH3, ankyrin repeats and pleckstrin domain (*ASAP1,* official name), is an ADP ribosylation factor GTPase-activating protein that interacts with signal transduction proteins involved in growth and differentiation, such as SRC, FAK, phosphatidylinositol 4,5-biphosphate, and CRK and regulates actin cytoskeletal remodeling that is necessary for cell motility. Overexpression of the ASAP1/DDEF1 protein disrupts focal adhesion turnover, thereby blocking cell spreading and promoting cell motility [29]. *DDEF1/ASAP1* expression is strongly correlated with *PTP4A3* ($R = 0.78$, $p < 0.001$), and biochemical interaction between these 2 factors remain to be studied.

MITF

MITF is a basic helix-loop-helix leucine zipper transcription factor involved in the maturation of neural crest-derived melanocytes, retinal pigment epithelium and bone marrow-derived mast cells and osteoclasts [66]. It is widely expressed in melanomas (cutaneous and uveal), but its role in melanocyte transformation is unclear. Over the years, MITF function in melanocyte homeostasis and melanoma development has been widely studied. In normal melanocytes, MITF dosage and activity acts as a rheostat to provide proproliferative or antiproliferative signals in order to promote an adequate cellular answer to external stimuli [67]. Several proteins that contribute to cell migration/invasion (DIAPH1 and c-MET), survival (BCL2, HIF1a, c-MET, and ML-IAP), and proliferation (CDK2) have been identified as MITF target genes [67]. MITF is critically required to prevent senescence and to favor melanoma cell proliferation [68].

MITF amplification occurs during skin melanoma progression, and ectopic *MITF* expression, in conjunction with the BRAF (V600E) mutant protein, transforms primary human melanocytes; thus, MITF can function as a melanoma oncogene [69]. However, a reduction in *MITF* expression has been reported to be linked to the metastatic behavior of the cells through its control of the *DIAPH1* gene [70], and *MITF* is located at 3p14.1, in a commonly lost chromosome in UM.

MicroRNAs

MicroRNAs (miRNAs) constitute a class of small (21- to 24-nucleotide) noncoding RNAs that reduce translation and stability of target mRNAs through sequence-specific interaction with their 3′-untranslated region. miRNAs play key roles in the development and establishment of cell identity [71, 72]. An aberrant expression or metabolism of miRNA has been linked to cancer. Abnormal levels of miRNA can affect many cellular processes that are routinely altered in cancer, such as differentiation, proliferation, apoptosis and migration. Some miRNAs exhibit differential expression levels in cancers and have demonstrated the capacity to affect cellular transformation and metastasis [73]. miR-137 targeting *MITF* is located at 1p22 [74], a region commonly lost in UM [20–22], and may therefore explain why deletion of chromosome 3 does not significantly decrease the *MITF* expression level. In addition, miRNAs have been identified as highly accurate biomarkers for metastatic risk in UM [75]. *MITF* expression is not correlated with PTP4A3 expression, but preliminary experiments suggest that PTP4A3 is able to modulate the MITF expression potential on target promoters (data not shown).

Biochemical Pathways Correlated with Metastasis Development

Epithelial-to-Mesenchymal Transition

Evidence accumulates that epithelial-to-mesenchymal transition (EMT) represents an important correlate of late-stage tumor progression, and recent reports indicate that the emergence of cells exhibiting stem-cell-like features occurs in part as a result of EMT, also contributing to drug resistance [76, 77]. EMT inducers, such as transforming growth factor-β or receptor tyrosine kinase ligands, trigger changes in gene expression by complex signaling networks. A paramount consequence of these signaling events is the upregulation of transcriptional repressors reducing E-cadherin (*CDH1* gene) expression. Such a reduction causes adherent-junction breakdown, and along with other signaling events such as modulation of Rho GTPase function, loss of cell polarity ensues [76, 77]. During EMT, cells acquire a spindle-shaped, highly motile fibroblastoid phenotype. This behavior differs from results showing increased expression of E-cadherin in class 2 tumors [39]. Experiments performed by Onken et al. [39] predict the existence of an activator associated with the E-box2 element of the *CDH1* promoter that is inhibited by ID2, a dominant-negative inhibitor of basic helix-loop-helix transcription factors [78]. *CDH1* is not differentially expressed between our *meta0* and *meta1* tumors, but *ID2* is significantly downregulated in *meta1* tumors, thus resulting in *CDH1* overexpression. *ID2* expression is inversely correlated with *PTP4A3* ($R = -0.71$, $p < 0.001$), suggesting a possibly negative interconnection between these 2 factors. ID2 also regulates the allocation of ectodermal stem cells into the neural crest lineage [79] and inhibits the expression of class 2 phenotypic features (e.g. epithelioid morphology, ECM patterns, and membranous E-cadherin expression).

Cancer Stem Cells

There is increasing evidence for the existence of cancer stem cells in many forms of cancer, and it appears that these cells are responsible for dissemination, repopulation, and recapitulation of the tumor at distant sites [80]. Indeed, class 2 tumors show significant transcriptional similarity to primitive neural and ectodermal stem cells [81]. In contrast, class 1 tumors show similarity only to more mature neural crest cells and differentiated melanocytes. Interestingly, class 2 tumors do not exhibit similarity to undifferentiated embryonic stem cells, suggesting that the class 2 signature does not correspond to a generalized 'dedifferentiation' but, rather, to the emergence of a lineage-specific primitive transcriptional program [81].

Conclusion

Characteristic DNA copy number and expression profiles have been identified in UM, which are strongly correlated with prognosis. Identification of the genetic pathways associated with metastasis is still in a preliminary stage, but several results suggest that distinct genes over- or underexpressed (through DNA copy gains or losses, or deregulated transcription) may be involved in cooperative pathways. Transcriptional deregulation is not the only way to deregulate biochemical pathways, since mutations have been found in genes involved in homologous reparation of double-strand DNA breaks (inactivating somatic mutations in *BAP1*) or modulating MAP kinase pathways (activating mutations in *BRAF* or *GNAQ*). It can be anticipated that new-generation sequencing performed on primary tumors, metastases, and constitutional DNAs from the same patient will disclose crucial newly mutated genes, shedding light on the mechanisms of UM tumorigenesis and metastasis development, and opening the way for rational therapies.

Acknowledgments

This work was supported by grants of the Department of Translational Research, Institut Curie, Cancéropole Ile-de-France, Association pour la Recherche sur le Cancer, Retina France, and Ligue Nationale Contre le Cancer.

References

1 Damato B: Does ocular treatment of uveal melanoma influence survival? Br J Cancer 2010;103:285–290.
2 Kilic E, van Gils W, Lodder E, Beverloo HB, van Til ME, Mooy CM, Paridaens D, de Klein A, Luyten GP: Clinical and cytogenetic analyses in uveal melanoma. Invest Ophthalmol Vis Sci 2006;47:3703–3707.
3 Landreville S, Agapova OA, Harbour JW: Emerging insights into the molecular pathogenesis of uveal melanoma. Future Oncol 2008;4:629–636.
4 Harbour JW: Molecular prognostic testing and individualized patient care in uveal melanoma. Am J Ophthalmol 2009;148:823–829.
5 Singh AD, Wang MX, Donoso LA, Shields CL, De Potter P, Shields JA: Genetic aspects of uveal melanoma: a brief review. Semin Oncol 1996;23:768–772.
6 Abdel-Rahman MH, Pilarski R, Ezzat S, Sexton J, Davidorf FH: Cancer family history characterization in an unselected cohort of 121 patients with uveal melanoma. Fam Cancer 2010;9:431–438.
7 Kannengiesser C, Avril MF, Spatz A, Laud K, Lenoir GM, Bressac-de-Paillerets B: CDKN2A as a uveal and cutaneous melanoma susceptibility gene. Genes Chromosomes Cancer 2003;38:265–268.
8 Hearle N, Damato BE, Humphreys J, Wixey J, Green H, Stone J, Easton DF, Houlston RS: Contribution of germline mutations in BRCA2, P16(INK4A), P14(ARF) and P15 to uveal melanoma. Invest Ophthalmol Vis Sci 2003;44:458–462.
9 Easton DF, Steele L, Fields P, Ormiston W, Averill D, Daly PA, McManus R, Neuhausen SL, Ford D, Wooster R, Cannon-Albright LA, Stratton MR, Goldgar DE: Cancer risks in two large breast cancer families linked to BRCA2 on chromosome 13q12–13. Am J Hum Genet 1997;61:120–128.
10 Iscovich J, Abdulrazik M, Cour C, Fischbein A, Pe'er J, Goldgar DE: Prevalence of the BRCA2 6174 del T mutation in Israeli uveal melanoma patients. Int J Cancer 2002;98:42–44.
11 Buecher B, Gauthier-Villars M, Desjardins L, Lumbroso-Le Rouic L, Levy C, De Pauw A, Bombled J, Tirapo C, Houdayer C, Bressac-de Paillerets B, Stoppa-Lyonnet D: Contribution of CDKN2A/P16 (INK4A), P14 (ARF), CDK4 and BRCA1/2 germline mutations in individuals with suspected genetic predisposition to uveal melanoma. Fam Cancer 2010;9:663–667.
12 Mudhar HS, Parsons MA, Sisley K, Rundle P, Singh A, Rennie IG: A critical appraisal of the prognostic and predictive factors for uveal malignant melanoma. Histopathology 2004;45:1–12.
13 Mensink HW, Vaarwater J, Kiliç E, Naus NC, Mooy N, Luyten G, Brüggenwirth HT, Paridaens D, de Klein A: Chromosome 3 intratumor heterogeneity in uveal melanoma. Invest Ophthalmol Vis Sci 2009; 50:500–504.
14 Onken MD, Worley LA, Tuscan MD, Harbour JW: An accurate, clinically feasible multi-gene expression assay for predicting metastasis in uveal melanoma. J Mol Diagn 2010;12:461–468.
15 Bronkhorst IHG, Maat W, Jordanova ES, Kroes WGM, Schalij-Delfos NE, Luyten GPM, Jager MJ: Effect of heterogeneous distribution of monosomy 3 on prognosis in Uveal Melanoma. Arch Pathol Lab Med, in press.
16 Onken MD, Worley LA, Person E, Char DH, Bowcock AM, Harbour JW: Loss of heterozygosity of chromosome 3 detected with single nucleotide polymorphisms is superior to monosomy 3 for predicting metastasis in uveal melanoma. Clin Cancer Res 2007;13:2923–2927.
17 Damato B, Dopierala J, Klaasen A, van Dijk M, Sibbring J, Coupland SE: Multiplex ligation-dependent probe amplification of uveal melanoma: correlation with metastatic death. Invest Ophthalmol Vis Sci 2009;50:3048–3055.
18 Young TA, Burgess BL, Rao NP, Gorin MB, Straatsma BR: High-density genome array is superior to fluorescence in-situ hybridization analysis of monosomy 3 in choroidal melanoma fine needle aspiration biopsy. Mol Vis 2007;13:2328–2333.
19 McCannel TA, Burgess BL, Rao NP, Nelson SF, Straatsma BR: Identification of candidate tumor oncogenes by integrative molecular analysis of choroidal melanoma fine-needle aspiration biopsy specimens. Arch Ophthalmol 2010;128:1170–1177.
20 Hughes S, Damato BE, Giddings I, Hiscott PS, Humphreys J, Houlston RS: Microarray comparative genomic hybridisation analysis of intraocular uveal melanomas identifies distinctive imbalances associated with loss of chromosome 3. Br J Cancer 2005;93:1191–1196.
21 Ehlers JP, Worley L, Onken MD, Harbour JW: Integrative genomic analysis of aneuploidy in uveal melanoma. Clin Cancer Res 2008;14:115–122.
22 Trolet J, Hupé P, Huon I, Lebigot I, Decraene C, Delattre O, Sastre-Garau X, Saule S, Thiéry JP, Plancher C, Asselain B, Desjardins L, Mariani P, Piperno-Neumann S, Barillot E, Couturier J: Genomic profiling and identification of high-risk uveal melanoma by array CGH analysis of primary tumors and liver metastases. Invest Ophthalmol Vis Sci 2009;50:2572–2580.

23 Onken MD, Worley LA, Ehlers JP, Harbour JW: Gene expression profiling in uveal melanoma reveals two molecular classes and predicts metastatic death. Cancer Res 2004;64:7205–7209.
24 Parrella P, Sidransky D, Merbs SL: Allelotype of posterior uveal melanoma: implications for a bifurcated tumor progression pathway. Cancer Res 1999;59:3032–3037.
25 Höglund M, Gisselsson D, Hansen GB, White VA, Säll T, Mitelman F, Horsman D: Dissecting karyotypic patterns in malignant melanomas: temporal clustering of losses and gains in melanoma karyotypic evolution. Int J Cancer 2004;108:57–65.
26 Tschentscher F, Prescher G, Horsman DE, White VA, Rieder H, Anastassiou G, Schilling H, Bornfeld N, Bartz-Schmidt KU, Horsthemke B, Lohmann DR, Zeschnigk M: Partial deletions of the long and short arm of chromosome 3 point to two tumor suppressor genes in uveal melanoma. Cancer Res 2001;61:3439–3442.
27 Parrella P, Fazio VM, Gallo AP, Sidransky D, Merbs SL: Fine mapping of chromosome 3 in uveal melanoma: identification of a minimal region of deletion on chromosomal arm 3p25.1-p25.2. Cancer Res 2003;63:8507–8510.
28 Cross NA, Ganesh A, Parpia M, Murray AK, Rennie IG, Sisley K: Multiple locations on chromosome 3 are the targets of specific deletions in uveal melanoma. Eye (Lond) 2006;20:476–481.
29 Ehlers JP, Worley L, Onken MD, Harbour JW: DDEF1 is located in an amplified region of chromosome 8q and is overexpressed in uveal melanoma. Clin Cancer Res 2005;11:3609–3613.
30 Onken MD, Worley LA, Harbour JW: A metastasis modifier locus on human chromosome 8p in uveal melanoma identified by integrative genomic analysis. Clin Cancer Res 2008;14:3737–3745.
31 Bennett DC: How to make a melanoma: what do we know of the primary clonal events? Pigment Cell Melanoma Res 2008;21:27–38.
32 Davies H, Bignell GR, Cox C, Stephens P, Edkins S, Clegg S, Teague J, Woffendin H, Garnett MJ, Bottomley W, Davis N, Dicks E, Ewing R, Floyd Y, Gray K, Hall S, Hawes R, Hughes J, Kosmidou V, Menzies A, Mould C, Parker A, Stevens C, Watt S, Hooper S, Wilson R, Jayatilake H, Gusterson BA, Cooper C, Shipley J, Hargrave D, Pritchard-Jones K, Maitland N, Chenevix-Trench G, Riggins GJ, Bigner DD, Palmieri G, Cossu A, Flanagan A, Nicholson A, Ho JW, Leung SY, Yuen ST, Weber BL, Seigler HF, Darrow TL, Paterson H, Marais R, Marshall CJ, Wooster R, Stratton MR, Futreal PA: Mutations of the BRAF gene in human cancer. Nature 2002;417:949–954.
33 Maat W, Kilic E, Luyten GP, de Klein A, Jager MJ, Gruis NA, Van der Velden PA: Pyrophosphorolysis detects B-RAF mutations in primary uveal melanoma. Invest Ophthalmol Vis Sci 2008;49:23–27.
34 Onken MD, Worley LA, Long MD, Duan S, Council ML, Bowcock AM, Harbour JW: Oncogenic mutations in GNAQ occur early in uveal melanoma. Invest Ophthalmol Vis Sci 2008;49:5230–5234.
35 Bauer J, Kilic E, Vaarwater J, Bastian BC, Garbe C, de Klein A: Oncogenic GNAQ mutations are not correlated with disease-free survival in uveal melanoma. Br J Cancer 2009;101:813–815.
36 Harbour JW, Onken MD, Roberson ED, Duan S, Cao L, Worley LA, Council ML, Matatall KA, Helms C, Bowcock AM: Frequent mutation of BAP1 in metastasizing uveal melanomas. Science 2010;330:1410–1413.
37 Tschentscher F, Husing J, Holter T, Kruse E, Dresen IG, Jöckel KH, Anastassiou G, Schilling H, Bornfeld N, Horsthemke B, Lohmann DR, Zeschnigk M: Tumor classification based on gene expression profiling shows that uveal melanomas with and without monosomy 3 represent two distinct entities. Cancer Res 2003;63:2578–2584.
38 Worley LA, Onken MD, Person E, Robirds D, Branson J, Char DH, Perry A, Harbour JW: Transcriptomic versus chromosomal prognostic markers and clinical outcome in uveal melanoma. Clin Cancer Res 2007;13:1466–1471.
39 Onken MD, Ehlers JP, Worley LA, Makita J, Yokota Y, Harbour JW: Functional gene expression analysis uncovers phenotypic switch in aggressive uveal melanomas. Cancer Res 2006;66:4602–4609.
40 Onken MD, Lin AY, Worley LA, Folberg R, Harbour JW: Association between microarray gene expression signature and extravascular matrix patterns in primary uveal melanomas. Am J Ophthalmol 2005;140:748–749.
41 Petrausch U, Martus P, Tonnies H, Bechrakis NE, Lenze D, Wansel S, Hummel M, Bornfeld N, Thiel E, Foerster MH, Keilholz U: Significance of gene expression analysis in uveal melanoma in comparison to standard risk factors for risk assessment of subsequent metastases. Eye 2008;22:997–1007.
42 Laurent C, Valet F, Planque N, Silveri L, Maacha S, Anezo O, Hupé P, Plancher C, Reyes C, Albaud B, Rapinat A, Gentien D, Couturier J, Sastre-Garau X, Desjardins L, Thiery JP, Roman-Roman S, Asselain B, Barillot E, Piperno-Neumann S, Saule S: High PTP4a3 phosphatase expression correlates with metastatic risk in uveal melanoma patients. Cancer Res 2011;71:666–674.

43 Matter WF, Estridge T, Zhang C, Belagaje R, Stancato L, Dixon J, Johnson B, Bloem L, Pickard T, Donaghue M, Acton S, Jeyaseelan R, Kadambi V, Vlahos CJ: Role of PRL-3, a human muscle-specific tyrosine phosphatase, in angiotensin-II signaling. Biochem Biophys Res Commun 2001;283:1061–1068.

44 Stephens BJ, Han H, Gokhale V, Von Hoff DD: PRL phosphatases as potential molecular targets in cancer. Mol Cancer Ther 2005;4:1653–1661.

45 Wang H, Vardy LA, Tan CP, Loo JM, Guo K, Li J, Lim SG, Zhou J, Chng WJ, Ng SB, Li HX, Zeng Q: PCBP1 suppresses the translation of metastasis-associated PRL-3 phosphatase. Cancer Cell 2010;18: 52–62.

46 Wu X, Zeng H, Zhang X, Zhao Y, Sha H, Ge X, Zhang M, Gao X, Xu Q: Phosphatase of regenerating liver-3 promotes motility and metastasis of mouse melanoma cells. Am J Pathol 2004;164:2039–2054.

47 Wang H, Quah YS, Dong JM, Manser E, Tang JP, Zeng Q: PRL-3 down-regulates PTEN expression and signals through PI3K to promote epithelial-mesenchymal transition. Cancer Res 2007;67:2922–2926.

48 Liang F, Liang J, Wang WQ, Sun JP, Udho E, Zhang ZY: PRL3 promotes cell invasion and proliferation by down-regulation of Csk leading to Src activation. J Biol Chem 2007;282:5413–5419.

49 Saha S, Bardelli A, Buckhaults P, Velculescu VE, Rago C, St Croix B, Romans KE, Choti MA, Lengauer C, Kinzler KW, Vogelstein B: A phosphatase associated with metastasis of colorectal cancer. Science 2001;294:1343–1346.

50 Miskad UA, Semba S, Kato H, Yokozaki H: Expression of PRL-3 phosphatase in human gastric carcinomas: close correlation with invasion and metastasis. Pathobiology 2004;71:176–184.

51 Parsons JT, Martin KH, Slack JK, Taylor JM, Weed SA: Focal adhesion kinase: a regulator of focal adhesion dynamics and cell movement. Oncogene 2000; 19:5606–5613.

52 Li Z, Zhan W, Wang Z, Zhu B, He Y, Peng J, Cai S, Ma J: Inhibition of PRL-3 gene expression in gastric cancer cell line SGC7901 via microRNA suppressed reduces peritoneal metastasis. Biochem Biophys Res Commun 2006;348:229–237.

53 Lee SB, Schramme A, Doberstein K, Dummer R, Abdel-Bakky MS, Keller S, Altevogt P, Oh ST, Reichrath J, Oxmann D, Pfeilschifter J, Mihic-Probst D, Gutwein P: ADAM10 is upregulated in melanoma metastasis compared with primary melanoma. J Invest Dermatol 2010;130:763–773.

54 Roy R, Wewer UM, Zurakowski D, Pories SE, Moses MA: ADAM 12 cleaves extracellular matrix proteins and correlates with cancer status and stage. J Biol Chem 2004;279:51323–51330.

55 Jalava SE, Porkka KP, Rauhala HE, Isotalo J, Tammela TL, Visakorpi T: TCEB1 promotes invasion of prostate cancer cells. Int J Cancer 2009;124: 95–102.

56 Cockman ME, Masson N, Mole DR, Jaakkola P, Chang GW, Clifford SC, Maher ER, Pugh CW, Ratcliffe PJ, Maxwell PH: Hypoxia inducible factor-alpha binding and ubiquitylation by the von Hippel-Lindau tumor suppressor protein. J Biol Chem 2000; 275:25733–25741.

57 Buscà R, Berra E, Gaggioli C, Khaled M, Bille K, Marchetti B, Thyss R, Fitsialos G, Larribère L, Bertolotto C, Virolle T, Barbry P, Pouysségur J, Ponzio G, Ballotti R: Hypoxia-inducible factor 1alpha is a new target of microphthalmia-associated transcription factor (MITF) in melanoma cells. J Cell Biol 2005;170:49–59.

58 Liang F, Luo Y, Dong Y, Walls CD, Liang J, Jiang HY, Sanford JR, Wek RC, Zhang ZY: Translational control of C-terminal Src kinase (Csk) expression by PRL3 phosphatase. Biol Chem 2008;283:10339–10346.

59 Lorsch JR, Dever TE: Molecular view of 43 S complex formation and start site selection in eukaryotic translation initiation. J Biol Chem 2010;285:21203–21207.

60 Cole PA, Shen K, Qiao Y, Wang D: Protein tyrosine kinases Src and Csk: a tail's tale. Curr Opin Chem Biol 2003;7:580–585.

61 Xu W, Harrison SC, Eck MJ: Three-dimensional structure of the tyrosine kinase c-Src. Nature 1997; 385:595–602.

62 Liang F, Liang J, Wang W-Q, Sun J-P, Udho E, Zhang Z-Y: PRL3 promotes cell invasion and proliferation by down-regulation of Csk leading to Src activation. J Biol Chem 2007;282:5413–5419.

63 Chaudhary A, Brugge JS, Cooper JA: Direct phosphorylation of focal adhesion kinase by c-Src: evidence using a modified nucleotide pocket kinase and ATP analog. Biochem Biophys Res Commun 2002;294:293–300.

64 Yeatman TJ: A renaissance for SRC. Nat Rev Cancer 2004;4:470–480

65 Maat W, el Filali M, Dirks-Mulder A, Luyten GP, Gruis NA, Desjardins L, Boender P, Jager MJ, van der Velden PA: Episodic Src activation in uveal melanoma revealed by kinase activity profiling. Br J Cancer 2009;101:312–319.

66 Steingrimsson E, Copeland NG, Jenkins NA: Melanocytes and the microphthalmia transcription factor network. Annu Rev Genet 2004;38:365–411.

67 Cheli Y, Ohanna M, Ballotti R, Bertolotto C: 15-year quest in search for MITF target genes. Pigment Cell Melanoma Res 2010;23:27–40.

68 Giuliano S, Cheli Y, Ohanna M, Bonet C, Beuret L, Bille K, Loubat A, Hofman V, Hofman P, Ponzio G, Bahadoran P, Ballotti R, Bertolotto C: Microphthalmia-associated transcription factor controls the DNA damage response and a lineage-specific senescence program in melanomas. Cancer Res 2010;70:3813–3822.

69 Garraway LA, Widlund HR, Rubin MA, Getz G, Berger AJ, Ramaswamy S, Beroukhim R, Milner DA, Granter SR, Du J, Lee C, Wagner SN, Li C, Golub TR, Rimm DL, Meyerson ML, Fisher DE, Sellers WR: Integrative genomic analyses identify MITF as a lineage survival oncogene amplified in malignant melanoma. Nature 2005;436:117–122.

70 Carreira S, Goodall J, Denat L, Rodriguez M, Nuciforo P, Hoek KS, Testori A, Larue L, Goding CR: Mitf regulation of Dia1 controls melanoma proliferation and invasiveness. Genes Dev 2006;20: 3426–3439.

71 Stadler BM, Ruohola-Baker H: Small RNAs: keeping stem cells in line. Cell 2008;132:563–566.

72 Stefani G, Slack FJ: Small non-coding RNAs in animal development. Nat Rev Mol Cell Biol 2008;9:219–230.

73 Ma L, Teruya-Feldstein J, Weinberg RA: Tumour invasion and metastasis initiated by microRNA-10b in breast cancer. Nature 2007;449:682–688.

74 Bemis LT, Chen R, Amato CM, Classen EH, Robinson SE, Coffey DG, Erickson PF, Shellman YG, Robinson WA: MicroRNA-137 targets microphthalmia-associated transcription factor in melanoma cell lines. Cancer Res 2008;68:1362–1368.

75 Worley LA, Long MD, Onken MD, Harbour JW: Micro-RNAs associated with metastasis in uveal melanoma identified by multiplexed microarray profiling. Melanoma Res 2008;18:184–190.

76 Singh A, Settleman J: EMT, cancer stem cells and drug resistance: an emerging axis of evil in the war on cancer. Oncogene 2010;29:4741–4751.

77 Thiery JP, Acloque H, Huang RY, Nieto MA: Epithelial-mesenchymal transitions in development and disease. Cell 2009;139:871–890.

78 Perk J, Iavarone A, Benezra R: Id family of helix-loop-helix proteins in cancer. Nat Rev Cancer 2005; 5:603–614.

79 Martinsen BJ, Bronner-Fraser M: Neural crest specification regulated by the helix-loop-helix repressor Id2. Science 1998;281:988–991.

80 Li F, Tiede B, Massague J, Kang Y: Beyond tumorigenesis: cancer stem cells in metastasis. Cell Res 2007;17:3–14.

81 Chang SH, Worley LA, Onken MD, Harbour JW: Prognostic biomarkers in uveal melanoma: evidence for a stem cell-like phenotype associated with metastasis. Melanoma Res 2008;18:191–200.

Simon Saule
Institut Curie, Centre Universitaire
Bâtiment 110
FR–91405 Orsay (France)
Tel. +33 1 69 86 71 53, E-Mail simon.saule@curie.u-psud.fr

Jager MJ, Desjardins L, Kivelä T, Damato BE (eds): Current Concepts in Uveal Melanoma.
Dev Ophthalmol. Basel, Karger, 2012, vol 49, pp 166–181

Therapeutic Options in Metastatic Uveal Melanoma

Pascale Mariani[a] · Vincent Servois[b] · Sophie Piperno-Neumann[c]

Departments of [a]Surgery, [b]Radiology, and [c]Medical Oncology, Institut Curie, Paris, France

Abstract

Systemic treatments for metastatic uveal melanoma patients give poor results. R0 resection of liver metastases showed a real benefit in survival in a very highly selected population. Based on our multivariate analysis, we propose surgical treatment to metastatic patients with time from diagnosis of uveal melanoma to liver metastases >24 months, number of liver metastases ≤4 lesions, and absence of detectable miliary disease. Liver MRI is currently the best imaging method in this context even if miliary disease is still difficult to diagnose. Molecular and chromosomal classification strongly predicting metastatic death has to be used to identify genetic profiles and pathways involved in the pathogenesis of uveal melanoma leading to new targeted therapeutic strategies.

The risk of developing metastases is the major challenge in uveal melanoma. The numerous advances in the treatment of the primary tumor have not resulted in any improvement in survival rates [1], and metastases are only rarely detected at the time of initial diagnosis (only in 5% of the patients).

Up to 50% of patients may develop metastases within a median time of 2.4 years [2, 3], leading to inevitable death because of potential micrometastases that are undetectable at the time of the primary tumor diagnosis.

Metastases occur via hematogenous spread (in contrast to cutaneous melanoma) in the absence of lymphatic drainage of the eye; liver is the predominant site of relapse in more than 80% of patients, and is not involved in only 10% of cases.

The Institut Curie series of 2,241 cases with a median follow-up of 6 years, reported by Desjardins et al. [4] in 2006, showed a metastasis rate of 19% for all tumors, with a 5-year survival rate of 78%.

Some cases of very late metastatic progression have been reported, up to more than 40 years after the diagnosis of the primary tumor [5]. Several hypotheses have been proposed to explain this supposed quiescence. It has been suggested that some metastases do not develop angiogenic activity, thereby limiting their growth. Alternatively, immune phenomena may play a decisive role [6]. It has been demonstrated that proliferation of tumor cells is actively controlled by CD8 memory T lymphocytes. The mechanism of action of CD4 and CD8 T lymphocytes that infiltrate metastases is currently under investigation [7].

Metastatic uveal melanoma has a very variable natural history, ranging from a fulminant course to very prolonged stable disease in the absence of any specific treatment. Although spontaneous prolonged stable disease is uncommon, it seriously complicates the interpretation of prolonged stable disease during treatment.

The recommended surveillance after treatment of the ocular tumor consists of ophthalmological examination and ultrasound of the eye every 6 months, as well as annual angiography to confirm local disease control. Hepatic ultrasound is recommended every 6 months for 10 years, as early diagnosis of limited liver metastases allows complete surgery of the pathological parts.

Overall survival (OS) is poor whatever the treatment [8], ranging from 2–6 months with best supportive care to 6–12 months with any systemic immunotherapy or chemotherapy. The only option today to improve survival is liver surgery, but R0 (microscopic complete) resection is performed in less than 25% of patients because of the presence of more extensive hepatic or extrahepatic disease. OS reaches 24 months in patients who underwent R0 liver surgery.

There is no standard adjuvant treatment to prevent (micro) metastases. However, adjuvant treatment may be more effective in the adjuvant setting, targeting micrometastases rather than disease with a large tumor burden.

Clinical risk factors for metastases and survival have been identified in the 1980s when enucleation was the standard local treatment for uveal melanoma. These factors include age at uveal melanoma diagnosis, largest tumor diameter and thickness, anterior location, ciliary body involvement, retinal detachment, extrascleral extension and histological cell type [4].

More recently, molecular biomarkers such as monosomy 3 and gene expression profile have been shown to be highly important. Monosomy 3 has been the gold standard for metastatic prediction in uveal melanoma [9, 10].

Today, genome-wide techniques of genomic and expression profiling make it possible to improve the characterization of high-risk uveal melanoma; in addition to monosomy 3, alterations on chromosomes 6 and 8q are linked to metastatic death [11]. The development of more precise techniques to detect chromosomal alterations such as FISH or array CGH led to the identification of partial deletions of chromosome 3 (isodisomy), association of monosomy 3 with an extralong arm of isochromosome 8q, losses of 8p and 16q and gain of 6p [12]. Monosomy 3 and gain of 8q are good predictors of metastatic risk in uveal melanoma patients.

Gene expression profiling has been shown to be a sensitive and specific prognostic marker. The gene expression signature from Onken et al. [13] identifies 2 molecular classes of uveal melanoma with low risk (class 1) and high risk (class 2) of metastases.

Table 1. Survival data for metastatic uveal melanoma patients

Reference	Number	Median OS months	MFI months	Factors correlated with survival
Diener-West et al. [14], 2005	739	3.7	32–42	NA
Bedikian, et al. [16], 1995	201	7	–	Serum APL, MFI
Gragoudas et al. [15], 1991	145	3.7	29	Age ≤60 years
Rietschel et al. [43], 2005	119	12.5	53	Lung/soft tissue/skin metastases only MFI Surgery/intrahepatic treatment Female Age ≤60 years

MFI = Metastasis-free interval; APL = alkaline phosphatase level.

Very few studies are dedicated specifically to metastatic uveal melanoma patients. Eighty publications from the last 30 years with data on the efficacy of treatment were identified by Augsburger et al. [8] in a recent paper. Only a third were prospective clinical trials (3 pilot studies, 2 phase I , 8 phase I/II and 15 phase II studies); none of them reported a randomized phase III trial. Only 12 described an unselected group of patients, and the median OS for the 2 largest series was 3.6 and 3.7 months, respectively. Survival was longer for smaller series of patients, whose metastases were identified by systematic testing, or in selected case series of patients, ranging from 5 to 24 months.

Few studies have investigated the characteristics of these patients by analyzing the sites of metastases, method of discovery of metastatic disease, treatment and outcome of therapy, factors that correlate independently with prolonged survival: age, gender, performance status, symptoms, site and size of the largest metastasis, metastasis-free interval and diagnostic procedure. Although prognostic factors for survival in the metastatic patients have not been clearly determined, rapidly progressive disease, impaired performance status (weight loss, fever, liver pain) and impaired liver function tests are known to have a negative impact.

The 4 most important series of metastatic patients are described in table 1.

Locoregional Treatment

When locoregional treatment is applied for uveal melanoma, the response rates are higher than with any systemic therapy, but the effect on survival remains uncertain.

Table 2. Chemoembolization in metastatic uveal melanoma patients

Reference	Cytotoxic	Number	RR, %	Median OS, months
Mavligit et al. [17], 1988	CDDP	30	46	11
Patel et al. [18], 2005	BCNU	24	21 (1 CR + 4 PR)	5
Vogl et al. [19], 2007	MMC	12	5 (3 PR + 5 SD)	21
Dayani et al. [20], 2009	CDDP/DOX/MMC	20	0 (13 SD)	9

RR = Response rate; CR = complete response; PR = partial response; SD = stable disease.

Chemoembolization

Chemoembolization using cisplatin provides response rates of 16–46% and survival rates between 6 and 10 months; this approach is also useful in the context of local salvage after failure of another therapy (table 2).

Immunoembolization

Several small series were reported for immunoembolization with lymphokine-activated killer cells or granulocyte-macrophage colony-stimulating factor (GM-CSF), however, with poor results [21].

Isolated Liver Perfusion

Isolated liver perfusion with melphalan and TNF has been performed in a very small number of patients with high rates of morbidity and toxicity; local control of liver disease is often accompanied by the appearance of extrahepatic metastases with no survival benefit [22]. In these studies, the vascular supply to the liver is isolated and infused via an extracorporeal circuit separated from the systemic circulation. The response rate reached 62%, and the median survival was around 1 year (table 3).

For all locoregional treatments, studies are not randomized, conducted in a limited number of patients, while treatment-related toxicities are frequent and can be severe. Progression at extrahepatic sites with control of hepatic metastases is also a major problem.

Table 3. Isolated hepatic perfusion studies

Reference	Number	RR, %	Median OS, months
Feldman et al. [22], 2004	22 (± TNF)	59	11
Feldman et al. [22], 2004	29	62	12
Van Iersel et al. [23], 2008	12	33	12
Van Etten et al. [24], 2009	8	37	11

RR = Response rate.

Table 4. Intra-arterial chemotherapy in uveal melanoma

Reference	Number	RR, %	Median OS, months
Leyvraz [25], 1997	31	40	14
Becker [29], 2000 + IL-2/IFN	23	22	12
Egerer [27], 2001	7	28	24
Siegel [28], 2006	16	28	22
Peters [26], 2006	101	36	15

RR = Response rate; ; IL-2/IFN = in this study, patients received subcutaneous interferon and interleukin 2.

Hepatic Intra-Arterial Chemotherapy

Uveal melanoma has a unique metastatic predilection for the liver. Direct infusion of chemotherapy into the hepatic artery increases the therapeutic index and decreases the systemic toxicity. The administration of the drug needs a catheter placed during laparotomy or radiological procedure by trained surgeons and radiologists, and connected to a subcutaneously implanted pump.

Thirty-one patients achieved a response rate of 40% in a single-center study [25] using fotemustine infusion over 4 h; the duration of response was 11 months and the mean survival was 14 months. Observed complications were abdominal pain (15%) and catheter thrombosis (30%).

A multicenter retrospective study with the same agent in 101 patients reported a response rate of 36% and a median survival of 15 months in this selected cohort [26].

A European multicenter prospective phase III study from the EORTC Melanoma group is currently ongoing in 240 patients with exclusively liver metastases, who are randomized to receive intravenous fotemustine versus hepatic intra-arterial fotemustine (table 4).

New Options in Regional Treatment

High-dose immunoembolization using GM-CSF versus BCNU in 53 patients demonstrated a promising OS of 20.4 months for patients treated with high-dose GM-CSF [30].

In a recent retrospective review of 11 pretreated patients [radioembolization with yttrium-90 (pure beta emitter)], Kennedy et al. [31] reported 1 complete and 6 partial responses in 9 evaluable patients; 80% of patients were alive after 1 year.

Systemic Treatment

Intravenous Chemotherapy

Intravenous chemotherapy with agents currently used for cutaneous melanoma (dacarbazine, cisplatin or nitrosourea) gives objective response rates of less than 20% and an OS of 4–9 months in monotherapy in retrospective series [32]. No survival benefit has been demonstrated in a randomized study compared to best supportive care.

In a retrospective series of 470 consecutive metastatic patients managed at the Institut Curie from 2000 to 2008, median OS was 13 months, being statistically significantly different to the OS of the first treatment in the metastatic setting: i.e., 28 months for surgery, 12 months for any systemic treatment, and 4 months for best supportive care [33]. Therapy was not randomized.

A German phase II randomized study of treosulfan combined with gemcitabine versus treosulfan alone in 48 patients showed a slight advantage of the combination, with a median progression-free survival of 3 months, and a 1-year progression-free survival of 18% [34].

Contribution of Cytokines

The contribution of cytokines alone or in combination remains controversial. A study using a combination of interleukin-2 and histamine is following the encouraging results reported with this combination in liver metastases from cutaneous or uveal melanoma [35].

Specific immunotherapy using a combination of tumor-specific antigens and differentiation antigens ± immunological adjuvant did not induce any specific immune response and did not modify patient survival, despite 20–30% of stable disease lasting more than 6 months [36].

Targeted Therapies

Targeted therapies address molecular abnormalities associated with tumor development and progression. Today, no specific cancer genes have been related to the pathogenesis of uveal melanoma, but potential targets involved in proliferation, apoptosis, invasion and metastasis, and angiogenesis have been identified, and clinical trials have been started for metastatic patients.

Targeted therapies can be beneficial for the metastatic patients in terms of survival or nonprogression rates in the absence of an objective tumor response. Stable disease may be associated with improved survival in this rare disease.

Targeted therapies also have to be tested in the adjuvant setting, after completion of the local treatment of the primary tumor, in high-risk patients identified by molecular and genomic prognostication.

The molecular pathogenesis of uveal melanoma differs from that of cutaneous melanoma, and specific approaches in cutaneous melanoma may not be relevant in uveal melanoma. Moreover, uveal melanoma patients are often excluded from cutaneous melanoma studies.

Preclinical studies suggest a potential benefit of receptor tyrosine kinases, inhibitors of Bcl-2, histone deacetylase, ubiquitin-proteasome, mitogen-activated protein kinase and phosphatidylinositol-3-kinase AKT pathways [37].

Two single-agent phase II trials testing imatinib mesylate included 20 patients and have been stopped because of disappointing results [38, 39]. The prognostic significance of KIT expression in uveal melanoma is not known, and KIT expression is not associated with a response to imatinib.

Targeting angiogenesis seems to be more promising in uveal melanoma. Intravitreal application of antiangiogenic agents is currently used to treat neovascular ocular diseases such as age-related macular degeneration or proliferative diabetic retinopathy.

Treatment with bevacizumab suppressed in vitro growth and in vivo hepatic establishment of micrometastases in experimental ocular melanoma [40].

Preclinical data suggest a potential clinical benefit of the combination of dacarbazine and antiangiogenic therapy.

A phase II study with temozolomide and bevacizumab as first-line treatment in metastatic patients is ongoing at the Institut Curie. Sunitinib and sorafenib are also being tested in combination schedules for uveal metastatic patients (www.clinicaltrials.gov).

Surgical Management of Liver Metastases

Hepatic resection should be considered for liver metastases of uveal melanoma because the liver is the first metastatic site and as a single organ involved in 60–80%

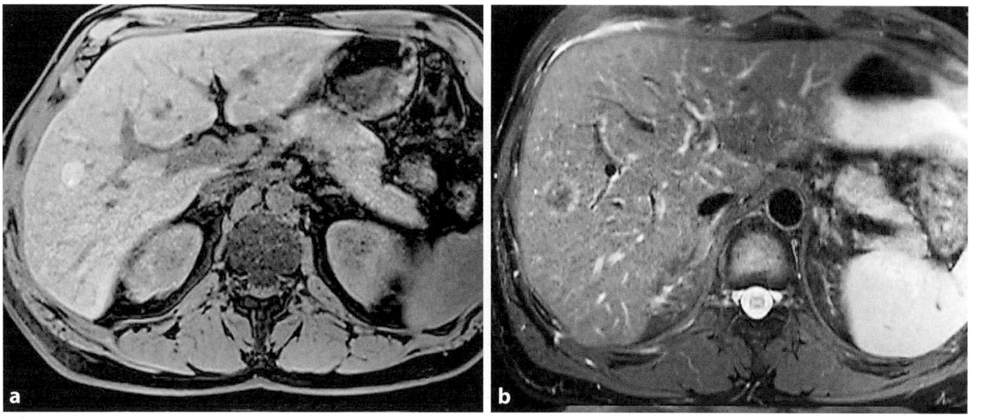

Fig. 1. Typical appearance of a metastatic melanoma: the center of the lesion containing high melanin concentration provides a high signal on T_1-weighted (**a**) and a low signal on T_2-weighted sequences (**b**).

of cases [41–43]. The best survival rates with established liver metastases have been obtained with maximal tumor reduction, where technically possible [44, 45].

Nearly 20 years ago, we developed an aggressive surgical liver approach at the Institut Curie because there was a lack of effectiveness of systemic treatments [44, 46].

Diagnosis of Metastases

After primary ocular treatment, patient monitoring combined clinical ophthalmological examination and abdominal ultrasonography every 6 months. If metastases were suspected, liver CT was performed to both confirm the diagnosis and evaluate the feasibility of resection. Lung metastases were excluded by chest CT and bone metastases with ^{99m}Tc bone scan.

Given its importance in the decision-making procedure, we will describe the technique of liver MRI. A special feature of liver metastases of uveal melanoma is to contain variable amounts of melanin pigment which will result in shortening of relaxation times T_1 and T_2, leading to a hyperintense lesion on T_1-weighted and a hypointense lesion on T_2-weighted sequences (fig. 1) [47, 48]. This characteristic appearance is not systematic and in our experience, at least one third of the lesions may show a signal similar to metastases from other primary tumors (T_1 hypointensity, T_2 hyperintensity). Our protocol includes a 2-dimensional fast spin echo T_2-weighted sequence with a mean echo time of 69 ms and respiratory triggering, a 2-dimensional fast spin echo T_1-weighted in- and out-phase sequence, and a 3-dimensional gradient echo T_1-weighted sequence before and

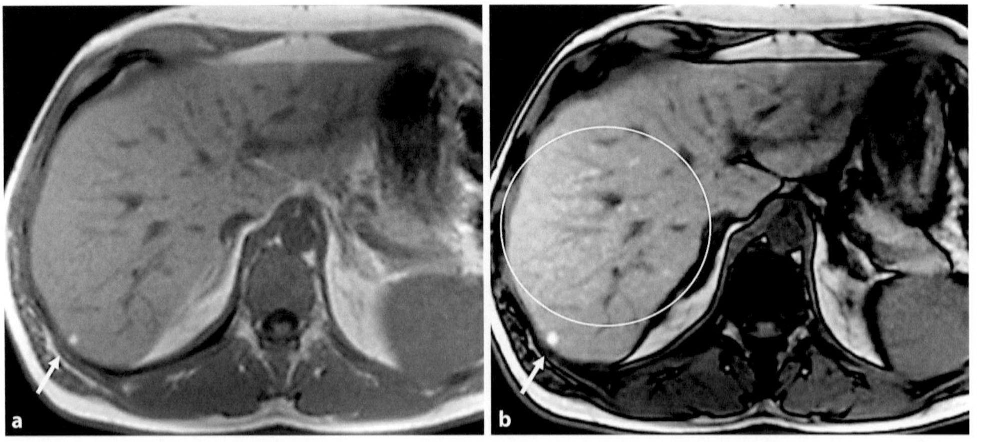

Fig. 2. T_1-weighted in-phase sequence (**a**) and out-phase sequence (**b**) of a small posterior subcapsular hyperintense lesion (arrows). Miliary intrahepatic disease (circle) is better depicted on the out-phase sequence.

after injection of gadolinium chelate with acquisition during the arterial (25–30 s), portal (60–70 s) and late (240 s) phases. The slice thickness is 6 mm for 2-dimensional sequences and 3.8 mm for the 3-dimensional sequence. In our recent series where the results were compared to surgery [49], the overall sensitivity of MRI was 67%, which is disappointing. It was still 100% for lesions of 5 mm or more in diameter. The most efficient sequences for the detection of small lesions were the 2-dimensional T_1-weighted out-phase and the 3-dimensional T_1-weighted arterial-phase dynamic acquisition (fig. 2). This result clearly reflects the original presentation of the metastatic disease that can involve macronodules and micronodules in the form of localized or diffuse miliary disease, which is often difficult to diagnose. One recent improvement in liver MRI is the introduction of diffusion-weighted sequence. Several studies have demonstrated the value of this sequence in the detection of metastatic lesions of colorectal origin [50]. We use a 6-mm slice thickness sequence with respiratory triggering and several b values (diffusion constant). Although it is still too early to conclude definitively, it seems that, as in other studies, a b value of 50 s/mm^2 is the most efficient. In some cases, this sequence allowed us to detect lesions which were not visible or difficult to detect on morphological sequences.

Selection of Patients for Hepatic Surgery

All patients with documented liver metastases were reviewed by the multidisciplinary team to assess the indication and suitability for surgical resection.

For the patients that underwent surgical intervention, the standard criteria for hepatic surgery were respected [age, performance status (WHO criteria), preservation of 30–40% normal hepatic parenchyma following complete metastasectomy]. The surgical objective was microscopically complete resection using either single or multiple resections. Surgery included intraoperative liver ultrasonography and frozen-section examination when a preoperative biopsy had not been obtained. The presence of 'miliary' metastases (multiple, diffuse, millimeter-sized, dark punctuate lesions) at laparotomy contraindicated major liver resection. Liver resections were classified as 'R0' (microscopically complete), 'R1' (microscopically incomplete) or 'R2' (macroscopically incomplete).

Our Experience at the Institut Curie

Between January 1991 and December 2009, 4,793 patients with uveal melanoma were treated at our institution for their primary tumor. Of these, 1,002 developed liver metastases and 317 underwent hepatic surgery. The median time between primary tumor diagnosis and hepatic surgery was 32 months (range, 0–216). Only 15 patients had liver metastases at the time of identification of the primary tumor.

During surgery, liver metastases were either solitary (n = 34), included 2–4 lesions (n = 71) or >4 lesions (n = 212). Among the 317 liver resections, there were 91 major liver resections (≥3 segments), 163 unisegmentectomies or metastasectomies and 63 liver biopsies. Since 2004, we performed 10 local radiofrequency ablations during laparotomy to attain an R0 resection. The following liver surgical resections were performed: R0 in 98, R1 in 25 and R2 in 194 cases. The presence of >4 macroscopic lesions during surgery was a strong predictor of miliary disease ($p = 0.00007$). The 30-day mortality rate was 2.8%, and the overall morbidity was 19%.

The median OS following hepatic surgery was 16 months [15–20]. The survival rate was 67.8% (± 2.7 SD) at 1 year and 7.0% (± 1.7 SD) at 5 years. The median OS was 23 months [20–24, 30–34] after an R0 resection, 15 months [10–24] after an R1 resection and 12 months [10–13] after an R2 resection. The median survival of 23 months in the R0 population compares favorably with intra-arterial chemotherapy results.

In univariate survival analysis (fig. 3), the parameters gender, age, tumor thickness and diameter, intraocular location of the primary melanoma and initial treatment were not predictive of outcome. A disease-free interval between primary tumor and liver metastases of more than 24 months was a significant prognostic factor of survival at 2 years ($p < 0.0001$); the number of liver metastases (≤4), their hepatic localization and the absence of miliary disease were all correlated with prolonged OS ($p < 0.0001$).

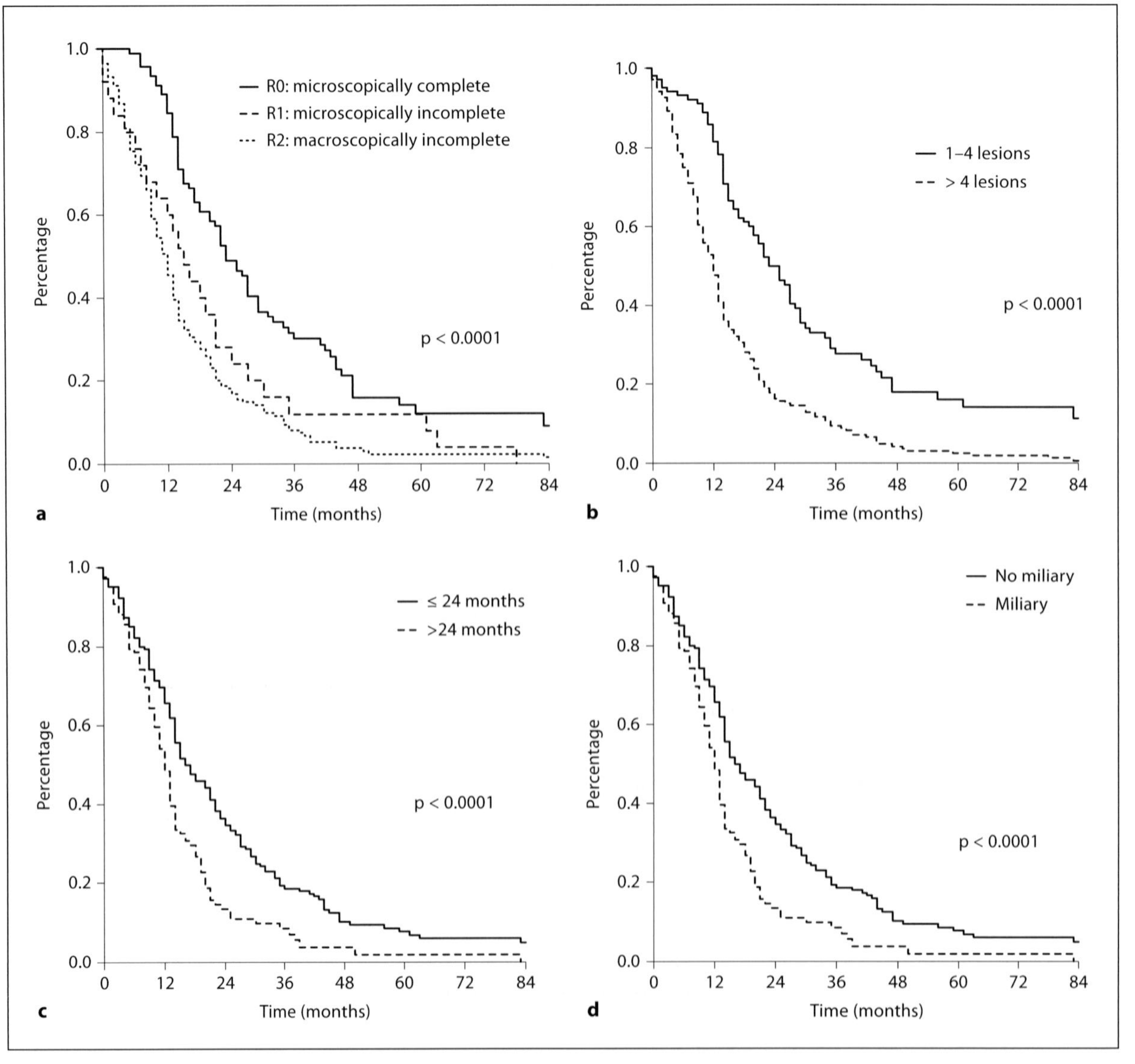

Fig. 3. OS after liver surgery. **a** OS according to quality of liver resection. **b** OS according to number of resected liver metastases. **c** OS according to time to liver metastases. **d** OS according to miliary disease.

Patient outcome was strongly correlated with the quality of liver resection, with the 2-year OS rate being 49% after R0 resection, 24% after R1 resection and 16.8% after R2 resection ($p < 0.0001$).

In multivariate survival analysis, we identified 3 variables that correlated independently with prolonged survival: time to liver metastases (>24 months), number of resected liver metastases (≤4 lesions,) and absence of miliary disease.

Other Surgical Series

Few teams have developed this surgical approach. Indeed, few teams regularly monitor their patients, and patients are not always sent to a liver surgery center.

Aoyama et al. [51] reported R0 resections in 9 patients (7 with liver metastases only and 2 with liver metastases and metastases at other sites). After R0 resection, the median recurrence-free and overall survival were 19 and 27 months, respectively. In a series of 24 patients, it was shown that the surgical cohort had a significantly longer median survival (38 months) than the nonsurgical cohort (9 months) [52]. However, of these, only 5 underwent liver resection, the remainder having other, nondefined metastasectomies. The R0 resection rate in our series, i.e. 98 of 317 (approximately 30%), was a little lower than that in the study by Rivoire et al. [45], being 50% (14 of 28) with similar semi-annual abdominal ultrasonography screening. The quality of surgery (R0 vs. R2) was the strongest indicator of improved outcome with 23 months of median survival for R0 being similar to the 25 months published previously [45].

R0 Resections Limited by Miliary Disease

As would be expected, with the number of metastases corresponding to disease 'load', fewer metastases correlated with a better survival. Specifically, the presence of less than 4 metastases was strongly significant ($p < 0.0001$) as was the absence of miliary disease ($p < 0.001$). Similarly, a unilobular distribution was much better than a multilobular distribution.

One very important finding in our series was that the presence of more than 4 metastases was strongly predictive for miliary disease in 82% of cases. Because diffuse miliary disease renders R0 resection impossible, this finding has changed our practice and we now perform diagnostic laparoscopy whenever more than 4 metastases are demonstrated or suspected on preoperative imaging. In this way, patients with miliary disease can be spared an unnecessary, and futile, laparotomy.

The Staging of the Intrahepatic Disease Is the Real Challenge

With respect to screening, liver function tests have been suggested to improve early diagnosis of liver metastases [53, 54]. Rietschel et al. [43], however, noted that blood tests were extremely inefficient for the detection of asymptomatic metastases. Only 1 patient in the study of Rivoire et al. [45] had liver function disruption with small metastases and, in our experience, liver function modification occurred only with massive, and therefore late, involvement.

To better enumerate the liver metastases, we opted for MRI because, apart from the specific characteristic of hyperintensity on T_1-weighting due to melanin, optimal contrast between the liver and the metastases appeared to allow an improved sensitivity of detection of small lesions. Another imaging technique commonly used in oncology is ^{18}FDG-PET scan. This whole body imaging that couples functional and morphological data has mainly been evaluated in the staging of cutaneous melanoma. There are few studies in uveal melanoma [55, 56]. In our series comparing preoperative FDG-PET scan to surgery [49], the overall sensitivity of FDG-PET scan for the detection of liver metastases was 44.7%; it was 61% for lesions greater than 5 mm in diameter. These results are equivalent to those of Ghanem et al. [57] for liver metastases from cutaneous melanoma which showed a sensitivity of 47%. They are probably related to technical factors and physiological factors unique to uveal melanoma. The main technical factors explaining the insufficient detection of lesions are the spatial resolution of PET, partial volume effects and errors in attenuation correction related to the mobility of the liver with respiration, and the level of a physiological high uptake by normal liver parenchyma. Some technical improvements such as respiratory gating and the late acquisition phase after FDG injection could improve results [58]. Some studies suggest that glucidic metabolism of uveal melanoma is variable [59]. In a recent study by Strobel et al. [60] comparing metastases of cutaneous and uveal melanoma, 59% of uveal melanoma lesions did not have a significant FDG uptake. All lesions but 1 were larger than 5 mm in diameter. These were mainly of the fusiform histological subtype. The lesions of the epithelioid subtype had a level of tracer uptake similar to cutaneous melanoma metastases. It is therefore possible that FDG is not the most efficient tracer and that other metabolic pathways using ^{11}C-labeled methionine or ^{18}F-DOPA would be more interesting in this context [61, 62].

Conclusion

Our results show a real benefit in survival for R0 resection of liver metastases (median OS 23 months). Patients have to be highly selected before surgery to offer the best results. Based on our multivariate analysis, we propose surgical treatment to metastatic patients with time from diagnosis of the uveal melanoma to liver metastases >24 months, number of liver metastases ≤4 lesions, and absence of detectable miliary disease.

Liver MRI is currently the best imaging method in this context even if miliary disease is still difficult to diagnose.

Systemic treatments for metastatic uveal melanoma patients give poor results. Molecular and chromosomal classification strongly predicting metastatic death has to be used to identify genetic profiles and pathways involved in the pathogenesis of uveal melanoma leading to new targeted therapeutic strategies.

References

1 Singh AD, Topham A: Survival rates with uveal melanoma in the United States: 1973–1997. Ophthalmology 2003;110:962–965.

2 Kujala E, Mäkitie T, Kivelä T: Very long-term prognosis of patients with malignant uveal melanoma. Ophthalmology 2003;110:962–965.

3 Desjardins L, Dorval T, Levy C, Cojean I, Schlienger P, Salmon RJ, Validire P, Asselain B: Etude randomisée de chimiothérapie adjuvante par le deticène dans le mélanome choroïdien. Ophtalmologie 1998; 12:168–173.

4 Desjardins L, Levy-Gabriel C, Lumbroso-Lerouic L, Sastre X, Dendale R, Couturier J, Piperno-Neumann S, Dorval T, Mariani P, Salmon R, Plancher C, Asselain B: Prognostic factors for malignant uveal melanoma. Retrospective study on 2,241 patients and recent contribution of monosomy-3 research. J Fr Ophtalmol 2006;29:741–749.

5 Abramson DH, Servodidio CA: Metastatic choroidal melanoma to the contralateral orbit 40 years after enucleation. Arch Ophthalmol 1997;115:134.

6 Schirrmacher V: T-cell immunity in the induction and maintenance of a tumour dormant state. Semin Cancer Biol 2001;11:285–295.

7 Blom DJ, Schurmans LR, De Waard-Siebinga I, De Volff-Rouendaal D, Keunen JE, Jager MJ: HLA expression in a primary uveal melanoma, its cell line, and four of its metastases. Br J Ophthalmol 1997;81:989–993.

8 Augsburger JJ, Corrêa ZM, Shaikh AH: Effectiveness of treatments for metastatic uveal melanoma. Am J Ophthalmol 2009;148:119–127.

9 Prescher G, Bornfeld N, Hirche H, Horsthemke B, Jöckel KH, Becher R: Prognostic implications of monosomy 3 in uveal melanoma. Lancet 1996;347: 1222–1225.

10 Onken MD, Worley LA, Ehlers JP, Harbour JW: Gene expression profiling in uveal melanoma reveals two molecular classes and predicts metastatic death. Cancer Res 2004;64:7205–7209.

11 Sisley K, Rennie IG, Parsons MA, Jacques R, Hammond DW, Bell SM, Potter AM, Rees RC: Abnormalities of chromosomes 3 and 8 in posterior uveal melanoma correlate with prognosis. Genes Chromosomes Cancer 1997;19:22–28.

12 Trolet J, Hupé P, Huon I, Lebigot I, Decraene C, Delattre O, Sastre-Garau X, Saule S, Thiéry JP, Plancher C, Asselain B, Desjardins L, Mariani P, Piperno-Neumann S, Barillot E, Couturier J: Genomic profiling and identification of high-risk uveal melanoma by array CGH analysis of primary tumors and liver metastases. Invest Ophthalmol Vis Sci 2009;50:2572–2580.

13 Onken MD, Worley LA, Person E, Char DH, Bowcock AM, Harbour JW: Loss of heterozygosity of chromosome 3 detected with single nucleotide polymorphisms is superior to monosomy 3 for predicting metastasis in uveal melanoma. Clin Cancer Res 2007;13:2923–2927.

14 Diener-West M, Reynolds SM, Agugliaro DJ, Caldwell R, Cumming K, Earle JD, Hawkins BS, Hayman JA, Jaiyesimi I, Jampol LM, Kirkwood JM, Koh WJ, Robertson DM, Shaw JM, Straatsma BR, Thoma J, Collaborative Ocular Melanoma Study Group: Development of metastatic disease after enrollment in the COMS trials for treatment of choroidal melanoma: Collaborative Ocular Melanoma Study Group Report No 26. Arch Ophthalmol 2005;123:1639–1643.

15 Gragoudas ES, Egan KM, Seddon JM, Glynn RJ, Walsh SM, Finn SM, Munzenrider JE, Spar MD: Survival of patients with metastases from uveal melanoma. Ophthalmology 1991;98:383–390.

16 Bedikian AY, Legha SS, Mavligit G, Carrasco CH, Khorana S, Plager C, Papadopoulos N, Benjamin RS: Treatment of uveal melanoma metastatic to the liver: a review of the M.D. Anderson Cancer Center experience and prognostic factors. Cancer 1995;76: 1665–1670.

17 Mavligit GM, Charnsangavej C, Carrasco CH, Patt YZ, Benjamin RS, Wallace S: Regression of ocular melanoma metastatic to the liver after hepatic arterial chemoembolization with cisplatin and polyvinyl sponge. JAMA 1988;260:974–976.

18 Patel K, Sullivan K, Berd D, Mastrangelo MJ, Shields CL, Shields JA, Sato T: Chemoembolization of the hepatic artery with BCNU for metastatic uveal melanoma: results of a phase II study. Melanoma Res 2005;15:297–304.

19 Vogl T, Eichler K, Zangos S, Herzog C, Hammerstingl R, Balzer J, Gholami A: Preliminary experience with transarterial chemoembolization (TACE) in liver metastases of uveal malignant melanoma: local tumor control and survival. J Cancer Res Clin Oncol 2007;133:177–184.

20 Dayani PN, Gould JE, Brown DB, Sharma KV, Linette GP, Harbour JW: Hepatic metastasis from uveal melanoma: angiographic pattern predictive of survival after hepatic arterial chemoembolization. Arch Ophthalmol 2009;127:628–632.

21 Keilholz U, Scheibenbogen C, Brado M, Georgi P, Maclachlan D, Brado B, Hunstein W: Regional adoptive immunotherapy with interleukin-2 and lymphokine-activated killer (LAK) cells for liver metastases. Eur J Cancer 1994;30A:103–105.

22 Feldman ED, Pingpank JF, Alexander HR Jr: Regional treatment options for patients with ocular melanoma metastatic to the liver. Ann Surg Oncol 2004;11:290–297.
23 Van Iersel LB, Hoekman EJ, Gelderblom H, Vahrmeijer AL, van Persijn van Meerten EL, Tijl FG, Hartgrink HH, Kuppen PJ, Nortier JW, Tollenaar RA, van de Velde CJ: Isolated hepatic perfusion with 200 mg melphalan for advanced noncolorectal liver metastases. Ann Surg Oncol 2008;15: 1891–1898.
24 Van Etten B, de Wilt JH, Brunstein F, Eggermont AM, Verhoef C: Isolated hypoxic hepatic perfusion with melphalan in patients with irresectable ocular melanoma metastases. Eur J Surg Oncol 2009;35: 539–545.
25 Leyvraz S, Spataro V, Bauer J, Pampallona S, Salmon R, Dorval T, Meuli R, Gillet M, Lejeune F, Zografos L: Treatment of ocular melanoma metastatic to the liver by hepatic arterial chemotherapy. J Clin Oncol 1997;15:2589–2595.
26 Peters S, Voelter V, Zografos L, Pampallona S, Popescu R, Gillet M, Bosshard W, Fiorentini G, Lotem M, Weitzen R, Keilholz U, Piperno-Neumann S, Stupp R, Leyvraz S: Intra-arterial hepatic fotemustine for the treatment of liver metastases from uveal melanoma: experience in 101 patients. Ann Oncol 2006;17:578–583.
27 Egerer G, Lehnert T, Max R, Naeher H, Keilholz U, Ho AD: Pilot study of hepatic intraarterial fotemustine chemotherapy for liver metastases from uveal melanoma. Int J Clin Oncol 2001;6:25–28.
28 Siegel R, Hauschild A, Kettelhack C, Kahler KC, Bembenek A, Schlag PM: Hepatic arterial Fotemustine chemotherapy in patients with liver metastases from cutaneous melanoma is as effective as in ocular melanoma. Eur J Surg Oncol 2007;33: 627–632.
29 Becker JC, Terheyden P, Kämpgen E, Wagner S, Neumann C, Schadendorf D, Steinmann A, Wittenberg G, Lieb W, Bröcker EB: Treatment of disseminated ocular melanoma with sequential fotemustine, interferon alpha, and interleukin 2. Br J Cancer 2002;87:840–845.
30 Yamamoto A, Chervoneva I, Sullivan KL, Eschelman DJ, Gonsalves CF, Mastrangelo MJ, Berd D, Shields JA, Shields CL, Terai M, Sato T: High-dose immunoembolization: survival benefit in patients with hepatic metastases from uveal melanoma. Radiology 2009;252:290–298.
31 Kennedy AS, Nutting C, Jakobs T, Cianni R, Notarianni E, Ofer A, Beny A, Dezarn WA: A first report of radioembolization for hepatic metastases from ocular melanoma. Cancer Invest 2009;27:682–690.
32 Zografos L: Tumeurs intraoculaires. Paris, Masson, 2002, pp 212–229.
33 Piperno-Neumann S, Chaaba H, Plancher C, Mariani P, Couturier J, Sastre-Garau X, Servois V, Asselain B, Desjardins L: Retrospective analysis of 470 metastatic uveal melanoma patients managed at Institut Curie (2000 and 2008). Proc ESMO. Ann Oncol 2010;21(suppl 8):viii401–viii407.
34 Schmittel A, Schmidt-Hieber M, Martus P, Bechrakis NE, Schuster R, Siehl JM, Foerster MH, Thiel E, Keilholz U: A randomized phase II trial of gemcitabine plus treosulfan versus treosulfan alone in patients with metastatic uveal melanoma. Ann Oncol 2006;17:1826.
35 Agarwala SS, Glaspy J, et al: Results from a randomized phase III study comparing combined treatment with histamine dihydrochloride plus interleukin-2 versus interleukin-2 alone in patients with metastatic melanoma. J Clin Oncol 2002;20:125–133.
36 Piperno-Neumann S: Therapeutic options in uveal metastatic melanoma patients. European uveal melanoma workshop, Paris, March 2007.
37 Triozzi P, Eng C, Singh AD: Targeted therapy for uveal melanoma. Cancer Treat Rev 2008;34:247–258.
38 Penel N, Delcambre C, Durando X, Clisant S, Hebbar M, Negrier S, Fournier C, Isambert N, Mascarelli F, Mouriaux F: O-Mel-Inib: a Cancéropôle Nord-Ouest multicenter phase II trial of high-dose imatinib mesylate in metastatic uveal melanoma. Invest New Drugs 2008;26:561–565.
39 Hofmann UB, Kauczok-Vetter CS, Houben R, Becker JC: Overexpression of the KIT/SCF in uveal melanoma does not translate into clinical efficacy of imatinib mesylate. Clin Cancer Res 2009;15:324–329.
40 Yang H, Jager MJ, Grossniklaus HE: Bevacizumab suppression of establishment of micrometastases in experimental ocular melanoma. Invest Ophthalmol Vis Sci 2010;51:2835–2842.
41 Rajpal S, Moore R, Karakousis CP: Survival in metastatic ocular melanoma. Cancer 1983;52:334–336.
42 Seregard S, Kock E: Prognostic indicators following enucleation for posterior uveal melanoma. A multivariate analysis of long-term survival with minimized loss to follow-up. Acta Ophthalmol Scand 1995;73:340–344.
43 Rietschel P, Panageas KS, Hanlon C, Patel A, Abramson DH, Chapman PB: Variates of survival in metastatic uveal melanoma. J Clin Oncol 2005;23: 8076–8080.
44 Salmon RJ, Levy C, Plancher C, Dorval T, Desjardins L, Leyvraz S, et al: Treatment of liver metastases from uveal melanoma by combined surgery-chemotherapy. Eur J Surg Oncol 1998;24:127–130.

45 Rivoire M, Kodjikian L, Baldo S, Kaemmerlen P, Négrier S, Grange JD: Treatment of liver metastases from uveal melanoma. Ann Surg Oncol 2005;12: 422–428.
46 Mariani P, Piperno-Neumann S, Servois V, Berry M, Dorval T, Plancher C, Couturier J, Levy-Gabriel C, Lumbroso-Le Rouic L, Desjardins L, Salmon RJ: Surgical management of liver metastases from uveal melanoma: 16 years' experience at the Institut Curie. Eur J Surg Oncol 2009; 35:1192–1197.
47 Prenikumar A, Marincola F, Taubenberger J, Chow C, Venzon D, Schwartzentruher D: Metastatic melanoma: correlation of MRI characteristics and histopathology. J Magn Res Imaging 1996;1:190–194.
48 Meli RJ, Ros PR: MR appearance of intra-abdominal metastatic melanoma. Magn Res Imaging 1992;10:7 05–708.
49 Servois V, Mariani P, Malhaire C, Petras S, Piperno-Neumann S, Plancher C, Levy C, Desjardins L, Salmon RJ: Comparison of MRI and FDG-PET for staging of uveal melanoma liver metastases. Eur J Surg Oncol 2010;36:189–194.
50 Nasu K, Kuroki Y, Nawano S, et al: Hepatic metastases: diffusion-weighted sensitivity-encoding versus SPIO-enhanced MR imaging. Radiology 2006;239: 122–130.
51 Aoyama T, Mastrangelo MJ, Berd D, Nathan FE, Shields CL, Shields JA, et al: Protracted survival after resection of metastatic uveal melanoma. Cancer 2000;89:1561–1568.
52 Hsueh EC, Essner R, Foshag LJ, Ye X, Wang He-Jing, Morton DL: Prolonged survival after complete resection of metastases from intra-ocular melanoma. Cancer 2004;100:122–129.
53 Diener-West M, Reynolds SM, Agugliaro DJ, Caldwell R, Cumming K, Earle JD, et al: Screening for metastases from choroidal melanoma: the collaborative ocular melanoma study group report 23. J Clin Oncol 2004;22:2438–2444.
54 Sato T, Babazono A, Shields JA, Shields CL, De Potter P, Mastrangelo MJ: Time to systemic metastases in patients with posterior uveal melanoma. Cancer Invest 1997;15:98–105.
55 Kurli M, Reddy S, Tena L, Pavlick A, Finger P: Whole body PET/CT staging of metastatic choroidal melanoma. Am J Ophthalmol 2005;140:193–199.
56 Francken A, Fulham M, Millward M, Thompson J: Detection of metastatic disease in patients with uveal melanoma using PET. Eur J Surg Oncol 2006;32:780–784.
57 Ghanem N, Altehoefer C, Högerie S, Nitzsche E, Lohrmann C, Schäfer O, Kotter E, Langer M: Detectability of liver metastases in malignant melanoma: prospective comparison of MRI and PET. Eur J Radiol 2005;54:264–270.
58 Dirisamer A, Halpern BS, Schima W, et al: Dual-time-point FDG-PET/CT for detection of hepatic metastases. Mol Imaging Biol 2008;10:335–340.
59 Lucignani G, Paganeli G, Modorati G, et al: MRI, antibody-guided scintigraphy and glucose metabolism in uveal melanoma. J Comput Assist Tomogr 1992;16:77–83.
60 Strobel K, Bode B, Dummer P, et al: Limited value of ^{18}F-FDG PET/CT and S-100B tumor marker in the detection of liver metastases from uveal melanoma compared to liver metastases of cutaneous melanoma. Eur J Nucl Med Mol Imaging 2009;36:1774–1782.
61 Lindholm P, Leskinen S, Nagren K, et al: Carbon-11-methionine PET imaging of malignant melanoma. J Nucl Med 1995;3:1806–1810.
62 Dimitrakopoulou-Strauss A, Strauss LG, Burger C: Quantitative PET studies in pretreated melanoma patients: a comparison of 6-(18F)fluoro-L-dopa with ^{18}F-FDG and (15)O-water using compartment and noncompartment analysis. J Nucl Med 2001;42: 248–256.

Pascale Mariani
Department of Surgery, Institut Curie
26, rue d'Ulm
FR–75248 Paris (France)
Tel. +1 44 32 46 45, E-Mail pascale.mariani@curie.net

Subject Index